Gliding for Gold

Gliding for Gold

THE PHYSICS OF WINTER SPORTS

Mark Denny

The Johns Hopkins University Press

BALTIMORE

© 2011 The Johns Hopkins University Press
All rights reserved. Published 2011
Printed in the United States of America on acid-free paper
9 8 7 6 5 4 3 2 1

The Johns Hopkins University Press
2715 North Charles Street
Baltimore, Maryland 21218-4363
www.press.jhu.edu

Library of Congress Cataloging-in-Publication Data

Denny, Mark, 1953–
    Gliding for gold : the physics of winter sports / Mark Denny.
       p. cm.
    Includes bibliographical references and index.
    ISBN-13: 978-1-4214-0214-7 (hardcover : alk. paper)
    ISBN-13: 978-1-4214-0215-4 (pbk. : alk. paper)
    ISBN-10: 1-4214-0214-9 (hardcover : alk. paper)
    ISBN-10: 1-4214-0215-7 (pbk. : alk. paper)
    1. Force and energy. 2. Sliding friction. 3. Winter sports. I. Title.
    QC73.D46 2011
    530—dc22
                                                        2011000459

A catalog record for this book is available from the British Library.

*Special discounts are available for bulk purchases of this book. For more information, please
contact Special Sales at 410-516-6936 or specialsales@press.jhu.edu.*

The Johns Hopkins University Press uses environmentally friendly book materials,
including recycled text paper that is composed of at least 30 percent post-consumer waste,
whenever possible.

CONTENTS

ACKNOWLEDGMENTS

I am grateful to Amanda Bird of the U.S. Bobsled and Skeleton Federation for permission to reproduce a few of the photos from their gallery; similarly, I am happy to acknowledge the help of Terry Kolesar of U.S.A. Curling, and Katie Perhai of the U.S. Ski and Snowboard Association in providing images. For allowing me to use their photos of skaters, I thank Adnan Hussain, Caroline Paré, and Brooke Novak. At the Johns Hopkins University Press, I am grateful to editor and keen winter sportsman (in his front yard) Trevor Lipscombe, art director Martha Sewall, and copyeditor extraordinaire Carolyn Moser.

Gliding for Gold

THE START LINES

Winter sports (and I mean sports played on snow or ice, not sports that happen to be played in winter, like football) are hugely popular. Ice hockey has been organized into professional leagues in many countries for decades; each team has thousands of fanatical supporters who follow every aspect of the game. Skiing is practiced as a pastime by millions; they and millions more watch professional skiers compete in high-profile events, such as the Winter Olympic Games. The 2002 Salt Lake City Games attracted 2.1 billion viewers worldwide, including 187 million in the United States. Though U.S. viewing figures were down for the 2006 Winter Olympics (to perhaps 176 million),[1] there is no doubt that both the familiar and the unusual sports that are contested at these games get a lot of attention.

Many of the less well-known Winter Olympic sports—for example, bobsled and related sledding events such as skeleton and luge—may pique your interest only once every four years, but anyone who watches these plucky athletes careening down an icy track at over 80 mph surely finds it exhilarating. It's more than exhilarating for the athletes, of course: "You'd be a looney if you weren't scared," Canadian luge competitor Chris Moffat said in 2006, after experiencing the fearsome Cesana track at the Turin games. "We are two inches above the ice going 130 kph in our underwear. It's not exactly the safest sport."[2]

1. Part of the reason for the fall in number of viewers was the inconvenient broadcast times in the United States for the live events; the 2006 Winter Olympic Games took place in Turin, Italy. Another reason might be the rise in popularity of TV reality shows, and in particular *American Idol*. Detailed analysis of viewing figures for the Salt Lake City games can be found in IOC (2002). Half the world (including 213 million Americans) watched the Vancouver 2010 games, pushing *American Idol* off the top of the ratings.

2. See Reuters (2006). A luge rider is known as a "luger" (LOO-jer).

My reaction, when I see these sleds pick up speed and swing around the bends, with the athletes more or less firmly attached, is to wonder about the physics involved. (How does the speed depend upon friction? How high up the bend does centrifugal force take them?) It's the same when I see a speed skater powering herself around a bend, leaning over at 45° with the fingers of one hand touching the ice, or when I see a ski jumper launch himself off a ramp. (How do skaters accelerate on such a slippery surface? How does their "stride length" vary around a curve? Why do ski jumpers always adopt a V-shape when airborne? How far can they jump?) If you have wondered about any of these questions, or many others that arise when thinking about the physics of winter sports (Why do curling rocks curl? How much do turns slow down a skier? Is the physics of snowboarding different from that of skiing? Why do most figure skaters spin counterclockwise?) then read on.

I have analyzed the dynamics of a number of winter sports. The math is summarized in technical notes at the end of the book, for those "mathletes" who are interested in the details. If you're not, but just want to understand the how and why without numbers, then stick to the main text. A lot of people are allergic to math—they would rather eat a pound of broccoli than ingest math—and yet they appreciate science and technology and want to learn about it. However, the only language that we have in common with Mother Nature is mathematics, and this can cause a problem. My job in writing this book is to act as interpreter: I have labored long and hard to translate the technical analyses (necessarily mathematical) into lucid and palatable nonmath explanations.

There are hundreds of books out there that seek to enlighten you about winter sports in general. Many of these books are intended for specialists (say, professional sportsmen or keen enthusiasts who want to refine their techniques, or who need to know about the biomechanics). Many other books are presented at a lower level that makes for easier reading but less enlightenment. Here, by way of contrast, I am aiming for a general understanding of the physical processes underlying winter sports, processes that act to influence the movement of athletes and of the equipment they use. My explanations are general but are not banal or trivial—you're getting the whole meal deal here, though packaged in a digestible format.

What kind of physics underlies winter sports? The motive power that propels an athlete toward the finish line is either muscle power or gravity (or both). Against these are the dissipative forces of sliding friction (say between snow and ski) and aerodynamic drag. Aerodynamic lift also plays a role in some of our sports. Much of the physics that I will investigate in the pages to follow will involve the interaction of these forces. Winter sports provide a "clean" application of the dissipative forces because they are so dominant and because they act in a fairly predictable manner. Contrast this with, say, the physics of summer sports. Sailing, to take one example, involves hydrodynamic as well as aerodynamic drag, and the drag coefficients of sailing boats vary significantly during a race because the sail shape changes (because of the wind action, or the helmsman's maneuvers). You can see that these extra complications make realistic analysis difficult. One of the appeals for a scientist to write a book explaining the physics of winter sports is the relative simplicity of the forces involved. So we anticipate pretty good predictions from our analyses. Believable predictions, in turn, tell us about the important parameters of winter sports. We will learn just how important it is to get off to a flying start in bobsled events; just how crucial aerodynamic lift, and weight, is to ski jumping success; just how sliding friction influences the skiers' technique. In this book, using physics to analyze winter sports will tell us more about both.

Units: if you choose to follow the math, you will find that I work with the metric system in my technical notes. The main text is more variable. In most cases I will retain metric units (they're good enough for Winter Olympics measurements, so they're good enough for us), but many English speakers still like feet and inches, pounds and ounces, so I may from time to time slip into these awkward but venerable units.

After an introductory chapter dealing with the slippery science of snow and ice, we delve into the science of snow sports and then ice sports. A hefty slice of this book follows in the wake of this delving—a metaphorical track left in the snow, if you like—and includes a section called Ponderables and the technical notes. Ponderables is a series of questions "for the interested reader" that arise from material covered in the main text. For some of these questions I provide hints to guide you toward the answer. The technical notes contain all the math analysis, gathered together so as to avoid breaking up the flow of the main text. The extensive bibliography includes primary and secondary sources, plus a number of Web references, including YouTube videos that demonstrate aspects of winter sports far better than can any written explanation.

1 SOLID WATER—
SPORTS AND SCIENCE

Welcome to the warm-up. In the next chapter you will be metaphorically hurled headfirst into the cold water of winter sports physics. By way of preparation, so that you may brace yourself, this chapter permits you to first dip a toe. Stated more mundanely (but also less alarmingly—I wouldn't want you to think that reading this book is going to be like taking a cold bath), I will spend a chapter describing these winter sports. More: you will need to know something about the ice and snow surfaces they are played on, and be brought up to speed about the frictional forces that act upon athletes and their equipment. So, first an overview of winter sports (those which are included in the Winter Olympic Games), second a description of the interesting and unusual physical properties of ice and snow, and third an introduction to the physics of sliding friction and of aerodynamic drag.

LET THE GAMES BEGIN

The Winter Olympic Games are a modern invention that began in Chamonix, France, in 1924. The last such games were held in Vancouver, Canada, in 2010, and the next will be in Sochi, Russia, in 2014. As we will see, not all of the sports began in Chamonix. Some—particularly the women's events—are much more recent. Here is the list of Winter Olympic sports that are played on ice:

- bobsled
- luge
- skeleton
- figure skating (including ice dancing)
- long-track speed skating

- short-track speed skating
- ice hockey
- curling

And on the fluffier form of solid water, these sports are played:

- alpine skiing
- snowboarding
- cross-country skiing
- ski jumping
- biathlon
- nordic combined

The 2010 Games were divided into 86 events (up from the original number of 16). Over half of them come from four sports: speed skating, cross-country skiing, alpine skiing, and biathlon.

Given that you are reading this book, it seems likely that you are familiar with one or two of these sports—perhaps you *sweep*, or *poptart*, or *hotdog*, or *axel*. Even so, the majority of winter sports may still be something of a mystery to you, and so, to clarify fuzzy notions, I will here provide a brief description of each.

Bobsled. Two-man, four-man, or two-woman bobsleds hurtle down artificial tracks that are typically 1,250 meters long (with about 15 bends). Bobsleds are steered. Competitors begin from a standing start and push their sled as much as 50 meters (hereafter abbreviated "m") before boarding. This sport has been part of the Winter Olympics since its inception and has been dominated by American, German, and Swiss athletes.

Luge. One or two men (or one woman) jump onto a sled feet first, on their backs, and steer down the bobsled track. As with bobsled and skeleton, multiple runs are timed to decide the winner. Germans and Italians have been particularly successful at luge events, which have been part of the Winter Olympics since 1964.

Skeleton. One athlete dives onto a small sled, which uses the same track as for bobsled events. The skeleton competitor lies prone, facing forward, and steers by body movement only. Skeleton sledding has been an Olympic event (for both men and women) only since 2002 and has been dominated by U.S. athletes.

Figure skating. Single skaters and pairs (including "mixed doubles" for ice dancing events) perform maneuvers—jumps, rotations, lifts, and combinations—of varying difficulty over a set time and are awarded points for their performance of each maneuver. An Olympic sport since 1924, figure skating has been dominated by Soviet and Russian skaters.

Long-track speed skating. Long-track skating competitions are timed events over set distances, around an oval track. Usually competitors start in pairs (at opposite ends of the oval, for team pursuit events). Distances vary from 500 to 10,000 m, although there is also a 40-kilometer (km) marathon (with a mass start). Very popular with the Dutch, speed skating has been an Olympic event for men since 1924; women's events did not appear until 1960.

Short-track speed skating. Short-track skating events are races (final position is all that matters) each with a mass start of four to six skaters, on a short oval track of circumference 110 m (365 ft). Distances raced are 500–1,500 m (always counterclockwise around the track), plus relay events of 5,000 m (men) and 3,000 m (women). This event was first part of the Olympics in 1992.

Ice hockey. The men's game dates back to Chamonix, but women (and professional National Hockey League players) had to wait until 1998 before they could compete at the Olympics. The rules of Olympic hockey are a little different from those of the NHL, but it's basically the same game. Canada and Russia dominate.

Curling. A strange and venerable sport, curling requires two teams, each of four players, to take turns sliding heavy granite rocks into a bull's-eye target. Closest to the center wins the "end." Ten such ends decide the game. Curling has been an occasional Olympic sport since the Winter Games started but has been a fixture only since 1998. Canada, Great Britain (in practice, Scotland), or Sweden usually win these events.

Alpine skiing. Alpine skiing is the downhill variety of skiing (with fixed heel bindings, in contrast to the hinged bindings of nordic skiing). It is another timed event, with medals awarded for the fastest times over one or two runs down a set course. *Mogul* events take place over bumpy terrain; *freestyle* involves getting airborne. Part of the Olympics since 1936, alpine events have been dominated by Austria, France, Italy, Switzerland, and the United States.

Snowboarding. Skiing with a single wide ski was introduced to the 1998 Winter Olympics in Nagano, Japan. *Halfpipe* events take place on a track shaped like half a cylinder, with steep sides. *Slalom* events, as in skiing, involve maneuvering around poles or through gates. Americans and Swiss have been particularly successful at Olympic snowboarding.

Cross-country skiing. Long-distance endurance competitions across relatively level ground, this nordic style of skiing has been an Olympic sport since 1924 for men (1952 for women). The medal tables usually feature Finns, Norwegians, Russians, and Swedes.

Ski jumping. The lunatic fringe competes in Olympic Games by launching themselves off of a slope and remaining airborne for over 100 m. Points are awarded for style as well as distance. Only male lunatics are allowed at the Winter Olympics (where they have been competing since its inception). Austrians, Finns, and Norwegians dominate.

Biathlon. A combination of cross-country skiing and rifle shooting (.22 cal), biathlon has been an Olympic event since the Squaw Valley, California, games of 1960. Germany, Norway, and Russia usually hit the target.

Nordic combined. A combo of cross-country and ski jumping, Nordic combined has been a men's event since 1924, dominated by Norwegians.

THE SCIENCE OF SOLID WATER

It may be a tautology to say that water dominates the Earth's oceans, but it is a less well-known fact that it also dominates the land, in solid form: 23% of the surface of our planet is covered in snow. Snow is water that freezes into crystals from a gaseous state in the atmosphere; ice is water that freezes from the liquid phase. Both forms of solid water are unusual and complex materials; their slipperiness is rare among solids and is not yet fully understood by scientists. Snow is a myriad of delicate ice crystal structures, and so snow is more complex even than ice: to the intricacies of ice crystal structure must be added the manner in which individual crystals combine. A sensible way for me to proceed, therefore, is to begin with a discussion of monolithic ice and then move on to its fluffier form.

"Whoa," you say. "I bought a book about winter sports—you don't need to tell me about ice and snow. I've been there; I know an ice cube when I've drunk one, and snow is snow. Get on with the sports and leave out the physics lecture—I know enough already about ice and snow." Ha! Read on.

Ice

There are many different ways in which liquid water molecules can arrange themselves to form a solid crystalline structure when circumstances dictate. These circumstances—which determine the *phase transition* from liquid to solid—depend mostly upon temperature and pressure, but other environmental factors (such as humidity, or the presence of dust or other contaminants) can influence the freezing process. Figure 1.1 shows the *phase diagram* for water, over the range of temperatures and pressures that are commonly found in nature. Phase transitions occur when a line is crossed. Thus, liquid water evaporates into steam, which condenses back to the liquid phase; water freezes to become ice, which melts back into water; ice sublimates to steam, which can turn into ice by deposition. The *triple point* of figure 1.1 is the combination of temperature and pressure at which water can exist in all three phases at once: solid, liquid, and gas. Water is the only common substance with a triple point that occurs at everyday temperatures and pressures.

Note that the ice of figure 1.1 is labeled *Ih*. We currently know of 15 forms of ice (numbered *I* to *XV*); more will be discovered, no doubt, before this book goes to press. The *h* in *Ih* stands for "hexagonal," which describes the organization of water molecules within the ice crystal. Almost all the ice that is found naturally on earth is of this form. High up in the atmosphere, where the temperatures and pressures are different from the ranges shown in figure 1.1, a cubic structure, *Ic*, is found.[1] At higher temperatures and pressures—different locations in the real estate of the phase diagram—other crystalline structures appear. These 15 (and counting) forms of ice have different physical properties. This multiplicity of structures is extreme: no other substance has so many known solid forms.

The lattice structure of *Ih* ice is known completely and is illustrated in figure 1.2. A single molecule of water, H_2O, consists of one oxygen atom and two hydrogen atoms, at an angle of 104.45° (fig. 1.2a). This angle is close to the angles that are found in two-dimensional hexagons (fig. 1.2b) and three-dimensional tetrahedra, so it is not surprising that ice in the form *Ih* can be viewed as sheets of water molecules arranged as hexagons, or as a repeated tetrahedral structure (fig. 1.2c).

---

1. *Ic* ice forms in the upper atmosphere by vapor deposition at temperatures below −130°C.

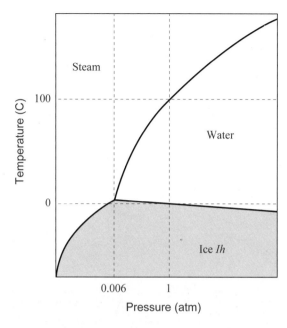

Figure 1.1.   Phase diagram for water at familiar temperatures and pressures.
Ice, the solid phase of water, can take at least 15 structural forms. By far the
most common form in nature is *Ih* ice.

Ih ice has a density of 917 kilograms per cubic meter (kg/m³) near 0°C,
whereas the density of water is 1,000 kg/m³. It is unusual for a chemical compound to have a solid form that is less dense than its liquid form.
In the case of water this phenomenon arises because of the crystalline
structure—there is a lot of space between molecules arranged in a hexagon. Ice density varies with temperature, as do other physical properties of
this solid (indeed, of most solids). In figure 1.3 you can see how the
density, thermal conductivity, and specific heat of *Ih* ice change as ice
temperature changes. (We will not need to know this level of detail about
the properties of ice in the chapters to follow; I include it here merely to
demonstrate the complexity of materials science and to show that crystalline structure influences the physical properties of a material. Different
forms of ice have different densities, for example.)
    Some of the mechanical properties of ice (for example, Young's modulus
—a measure of elasticity and, in particular, of shear stress) vary with
direction—again, a consequence of the crystalline structure. There exist
planes of weakness within the ice, so that it shears more easily in one plane

than in another plane. For us, the most important property of ice happens at the surface: it is slippery. Slipperiness is unusual in solid materials, and the slipperiness of ice particularly so—indeed, it is not yet fully understood. I will get to grips, so to speak, with this subject in the next section. Material scientists think that part of the slipperiness is, like shear stress, due to the crystalline structure of *Ih* ice. At surfaces the crystal structure

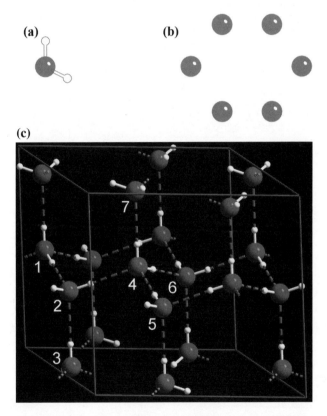

Figure 1.2. Hexagonal structure of ice. (a) A single water molecule consisting of one oxygen atom (large sphere) and two hydrogen atoms (small loops). (b) In *Ih* ice the oxygen atoms of six water molecules arrange themselves in a hexagon. The hydrogen atoms (not shown) are arranged differently in different forms of ice. (c) The 3-D structure of *Ih* ice. We can see the hexagonal component (or half of one) in molecules 1, 2, and 3, for example. Molecules 2, 5, 6, and 7 form a tetrahedron, with molecule 4 at its center. There are lots of gaps in this structure, which means that ice has a lower density than liquid water. The solid white lines and dashed gray lines show different types of chemical bond. Adapted from a Wikipedia figure.

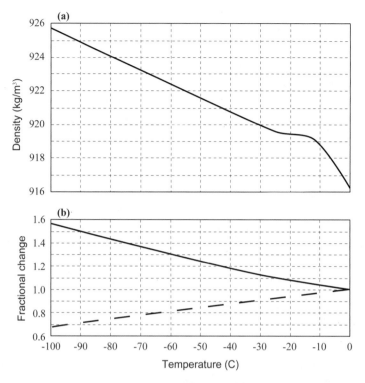

Figure 1.3. Three physical properties of *Ih* ice: (a) the density of *Ih* ice as a function of temperature; (b) fractional change in *Ih* ice thermal conductivity (solid line) and specific heat (dashed line) vs. temperature.

leaves many dangling, broken chemical bonds that give rise to a liquid-like behavior which accounts for certain familiar yet unique properties, including sintering and regelation as well as slipperiness. *Sintering* is the welding together of ice crystals without melting, as when loose snow combines to form a snowball. *Regelation* is the property of melting under pressure and freezing when the pressure is released. This property is often demonstrated to physics and engineering students as shown in figure 1.4.[2]

Snow

Water vapor crystallizes around dust particles in the atmosphere when the temperature is low enough and the humidity high enough. These crystals

2. For a readable, popular account of the structure and properties of *Ih* ice see Chang (2006). For technical details, see Schulson (1999).

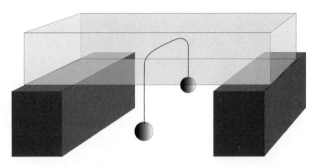

Figure 1.4. This classroom demonstration shows a strange property of ice—regelation—in action. Two weights are suspended from a thin wire that has been draped over a block of ice. Pressure melts the ice immediately beneath the wire, so that it cuts through the block. But once the pressure is released, the ice refreezes, so the block remains intact. Thus, the wire passes right through the block of ice without breaking it.

grow in a myriad of hexagonal patterns, forming snowflakes (fig. 1.5); the snowflakes quickly become too large to float around in the air and so they fall to the ground. There are no fewer than 10 forms of frozen precipitation, of which 7 are snow.[3] The shape of snowflakes is very sensitive to the meteorological conditions in the atmosphere where they form. Hexagonal ice crystals can aggregate into stars and plates, or into needles and columns. Stars and plates predominate when the air temperature is $-2°C$; columns and needles are produced more often when the temperature falls to $-5°C$; stars and plates again dominate at $-15°C$; at $-30°C$ all types are common. More complex shapes result if humidity is very high.[4]

Whatever shape the snowflakes choose for themselves, when they hit the ground they add to the *snowpack*. Snow on the ground is often layered; each day the top level melts and then freezes at night, producing a crust of ice, which is then covered by another layer of snow during the next snowfall. Snowpack layers may consist of different types of snow with different

3. The other three types of frozen precipitation are *hail*, *sleet*, and *graupel*. Hail is frozen raindrops; sleet consists of smaller particles that are part frozen and part liquid. Graupel is snow that has accumulated a lot of *rime*. Rime is the white ice that forms when water droplets freeze to an object. If the object happens to be a snowflake, the rime growth results in graupel.

4. See Kenneth Libbrecht's Web site, www.snowcrystals.com, or the Microsoft *Encarta* article on snow (Microsoft 2005), for popular accounts of snow formation. For a detailed technical account see Libbrecht (2005).

Figure 1.5. Photographs of snowflakes. Plate XIX of Wilson Bentley, "Studies among Snow Crystals," in *Annual Summary of the Monthly Weather Review*, 1902. Bentley, from Vermont, was the first person to photograph snowflakes.

properties; there is often a temperature gradient, which causes the snow crystals to evolve differently. A low gradient, for example, often leads to a rounding of the crystals, which can then compress more tightly, producing a stable snowpack. High temperature gradients, on the other hand, lead to the snow crystals' developing extra facets known as *depth hoar*.[5] Crystals with depth hoar do not bond well, and the resulting snowpack is often unstable.

For physicists who are trying to understand the effects of snow upon winter sports—say, for example, we want to learn about the friction that is generated between a ski and the snowpack surface—snow variability is a complicating factor. Snow can be hard or soft, compacted or fluffy, with a high or low water content. All of these factors and more (ambient temperature and humidity, ski material and shape, ski wax, even ski color) influence the ease with which a ski glides over snow, as we will see. To some extent, course preparation prior to an event reduces the variability (by compacting very loose snow, for example), but even so, athletes who do their thing over snow have to contend with a more capricious surface than athletes who glide over ice.

DISSIPATION

Dissipation, to a physicist, does not mean getting drunk on a Saturday night (though many physicists that I know are happy to do so). Instead, the word applies to forces that spread energy around. Friction is a dissipative force that saps energy from a moving object, such as an athlete in a time trial. Three types of friction contribute to the physics of winter sports, and these arise from the three phases of matter—solid, liquid, and gas, as illustrated in figure 1.6. Friction between two solid objects is called *sliding friction* (also known as *dry* or *contact friction*) whereas friction between two solids that are separated by a thin liquid layer is here called *lubricated friction*. Friction generated by a solid object (such as a hockey puck) passing through the air is called *aerodynamic drag*. The physics of these three forms is quite different, as we will now see.

At first glance, this subject may seem to be about as exciting as watching paint dry or filling out your tax return. If so, then all I can say by way of

---

5. *Hoar* refers to lightly feathered crystals that arise from the freezing of humid cold air. This can happen as water rises through the snowpack and evaporates at the surface, and then freezes to the surface layer.

(a)

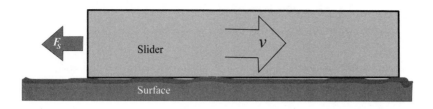

(b)

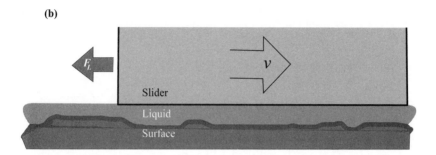

(c)

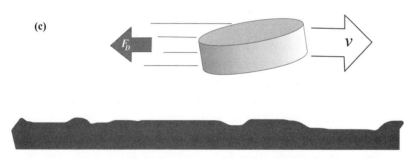

Figure 1.6. Three forms of friction force. (a) *Sliding friction*, F_S, involves one solid sliding over the surface of another solid. (b) *Lubricated friction*, F_L, arises between two solid surfaces that are separated by a thin layer of liquid. (c) The friction felt by a solid object, such as this hockey puck, flying through the air is called *aerodynamic drag*, F_D. In all cases, the force of friction is directed opposite to the velocity direction of the moving object, which moves from left to right.

encouragement is this: *suck it up*. Friction is absolutely crucial to understanding the physics of winter sports, and you need to know about it. So there.

Slip Sliding Away

The physics of sliding friction was unraveled in the seventeenth and eighteenth centuries by two Frenchmen, Guillaume Amonton and Charles-Augustin de Coulomb. (Earlier, Leonardo da Vinci had made some inroads into the problem.) Details are provided in technical note 1 for readers who are following the math; for those who are not, here is a non-mathematical explanation. Experiments with blocks of wood and other materials sliding over each other—say, wood sliding on stone, or glass sliding on metal—show that the force required to drag the block at constant speed depends upon the weight of the block. That is, for a wide range of block speeds (relative to the surface the block is sliding over), and for a variety of block shapes and sizes, force is proportional to block weight. The constant of proportionality is called the *coefficient of kinetic friction*, usually denoted by the Greek letter μ (mu), and is different for different block and surface materials. Thus, for example, when a wooden block is dragged over stone at constant speed, the coefficient of friction is about 0.4. So, to drag the block we must exert a force that is 40% of the block weight. This force is required to overcome the sliding friction between these two surfaces. For wood sliding over wood, the friction coefficient is 0.3; for copper sliding on glass it is 0.53.

We specify constant speed when defining a friction coefficient because at constant speed there can be no net force acting on the slider, according to Newton's first law. This means that the force of friction must be exactly cancelled by the dragging force. So our well-established experimental result (dragging force is proportional to slider weight) can be restated as *frictional force is proportional to slider weight*.

There are a couple of complications that make sliding friction a bit more involved than my description suggests. One of these complications arises when the slider is moving over a surface that is not level—such as when a block is sliding down an inclined plane. In this case the friction force is not proportional to block weight but to the component of weight that is perpendicular to the surface of the incline. I fully address this complication in technical note 1 because it will be important for us later, when we need to calculate the friction force of a skier or a bobsled traveling down an icy slope. Another complication is the friction force that applies when the block is not moving. For many materials, it seems to be the case that we need to exert more effort to get the block moving than we require to keep it moving. You may have experienced this phenomenon yourself—say when moving a heavy piece of furniture across a room. It takes a lot of effort to start the furniture moving, but less to maintain the movement; this is why we resort to an initial hard shove or jerk. Physicists describe this phenomenon by assigning a different value for the friction coefficient of stationary objects. For example, the *coefficient of static friction* for wood on stone is about 0.5; for wood on wood it is the same; for copper on glass it is about 0.68. Note that these values are larger than the corresponding kinetic friction coefficients.

Viewed under a magnifying glass, the phenomenon of sliding friction looks like the schematic illustration of figure 1.7. The caption explains why, for most solid sliders, the friction force is more or less independent of slider area (that is, the area of the block in fig. 1.6a, viewed from above). Viewed through a microscope the situation is more complicated. Two solid surfaces in contact form chemical bonds that are quickly broken when the surfaces slide over each other. This making and then breaking of bonds is the ultimate cause of sliding friction. We do not need to go there, however; for a reasonable understanding of the effects of sliding friction upon winter sports, all we need is the macroscopic view, that friction force equals friction coefficient multiplied by slider weight.[6]

Oiling the Wheels

If one solid object is sliding over another, and they are separated by a thin layer of liquid, the liquid can act as a lubricant that reduces the force of friction acting upon the slider. The physics of hydrodynamic lubrication, to give the subject its technical title, was unraveled in the late nineteenth century by Nikolai Pavlovich Petrov and Beauchamp Tower, among others. Lubrication works only for certain solids and liquids, and it usually requires the two surfaces to be quite smooth. Thus, metal machine parts benefit from oiling because the oil film significantly reduces the friction

6. Any undergraduate physics textbook will tell you about sliding friction; see, for example, Symon (1960, chap. 1) or Lazar (2003, chap. 3). For an account of sliding friction on ice, in the context of ice hockey, see Haché (2002).

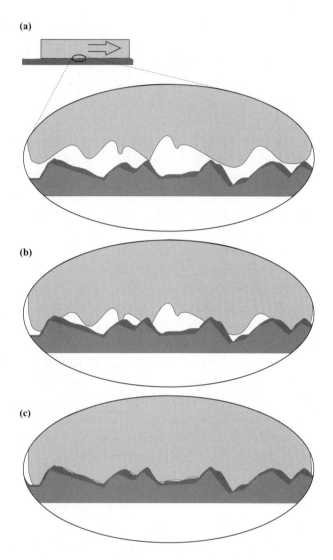

Figure 1.7. Sliding friction on the small scale. (a) For two hard surfaces, the points of contact are very few, so that the area of contact is much less than the geometrical area of the slider. In practice this means that the kinetic friction coefficient does not depend much upon the geometrical area. (b) If the slider weight increases, then the slider is pressed into the underlying surface, thus increasing the contact area and so the friction force. (c) For a slider material that is elastic and easily deformed, such as rubber, the slider takes on the shape of the underlying surface. For such materials, unlike hard substances, the force of friction *does* increase as area increases. This is why car tires are made wider to increase grip.

between the metal surfaces, and thus also reduces machine wear and heat generation. Car tires may grip a road surface tightly, but if the tires are slick and the road surface smooth, and if it is raining heavily, then hydroplaning results; here, we have an example of water providing a lubricating layer. In general, providing lubrication between two solids that are sliding past one another reduces the force of friction by a factor of 2 to 4.

Given that water can act as a lubricant, we need to know about lubricated friction, because water layers abound in winter sports. The physical setting is easily sketched (see fig. 1.6b). We can readily see why lubricated friction is less than sliding friction: the force of friction is reduced when the two surfaces are separated by a lubricating layer. Molecules of liquid are free to slide over one another—it is a defining property of liquids. Technically we say that liquids cannot support a *shear force*, unlike solids. You can see that the lubricating layer of figure 1.6b is being sheared; molecules of liquid near the slider are dragged to the right by the slider, whereas molecules of liquid near the stationary surface are hardly moving. In between there is a velocity gradient. The liquid layer is being smeared like butter on bread; this is what we call *shear*.

Many experiments have shown how lubricated friction force depends upon the physical parameters of the solids and of the liquid layer. Theory can predict the observed behavior. We know that the friction force acting upon the slider increases with the geometrical area of the slider and increases with slider speed. Friction is reduced as the thickness of the liquid layer increases. There is a constant, analogous to the coefficient of sliding friction, that relates these parameters (slider area and speed, and liquid layer thickness). It is called the *coefficient of dynamic viscosity* and is discussed, along with the mathematical expression for lubricated friction force, in technical note 1.[7]

Thus lubricated friction acts differently than sliding friction. Lubricated friction depends upon slider speed and area, but sliding friction does not. Sliding friction depends upon slider weight, but lubricated friction does not.

---

7. Lubricated friction is discussed in many fluid mechanics textbooks, such as Douglas and Matthews (1996, chap. 12) or Massey (1989, chap. 6). For a historical review of the subject, see Pinkus (1987).

What a Drag

Aerodynamic drag, like all of fluid dynamics, is a complex subject. If we were investigating the flight of a jet fighter or the ballistic trajectory of a bullet, then we would see a great deal of this complexity. However, we will be dealing with relatively low-speed projectiles (skiers—those who hurtle down slopes and those who jump—and skaters), and so the physics is more tractable. Some aspects of this subject are still not understood today and are the focus of active research, but the basic physics of aerodynamic drag became known in the early twentieth century. (Developing the theory of aerodynamic lift and drag was largely an achievement of German physicists, such as Ludwig Prandtl and Albert Betz.)

Newton realized, 350 years ago, that the force of air resistance—aerodynamic drag—depends upon the speed of the projectile. (As you've already noticed above, I am referring to the object that is subjected to drag force—whether a skydiver, a Scud missile, or a ski jumper—as a *projectile.*) In fact, drag force increases as the square of projectile speed, so if a projectile accelerates to three times its previous speed, the drag force becomes nine times greater. Also, the drag force increases as the cross-sectional area of the projectile (that is, the area presented to the wind—the area as seen from in front). The only other parameter that enters into the equation is air density. Drag force increases as air density increases. For our purposes, we can assume that air density is constant (there is only a small difference between air density at the top of a ski slope and at the bottom, for example).

These parameters are connected to drag force via the *drag coefficient.* This coefficient is the equivalent of sliding friction coefficient, and it depends upon projectile shape. In general, the drag coefficient varies with two fluid dynamical factors called the *Reynolds number* and the *Mach number,* but because our winter sports projectiles are low-speed, we can ignore this dependence, most fortunately, and can consider the drag coefficient to be a constant. In principle, the drag coefficient of a skier changes whenever he or she changes stance—say, from an upright position to a more aerodynamic crouch—but here I will assume a constant value. All our winter sports projectiles will be assigned a representative, average drag coefficient which will vary throughout a race as the racer—the projectile—changes stance but will not vary by much. Stated more precisely: I will make the reasonable assumption that the movement of an athlete around a winter sports track can be described using their average value for drag coefficient almost as accurately as if I were describing the movement using the drag coefficient instantaneous value.

Clearly, winter sports athletes these days are well aware of the importance of aerodynamic drag. They do their utmost to minimize their drag coefficient—for example, by wearing tight-fitting and smooth outer garments (just look at a speed skater: Spiderman on ice). Much effort, both theoretical and experimental, has been put into the design of bobsleds, to minimize their aerodynamic drag (always subject, of course, to the stringent design constraints of the sport's rules and regulations).

The math governing drag force is outlined in technical note 1.[8] Again, we have different behavior than obtains for the other types of friction. Aerodynamic drag is proportional to the square of projectile speed, whereas lubricated friction is proportional to speed, and sliding friction is independent of the speed. The mix of these three dissipative friction forces makes the physics of winter sports interesting. We have seen already that the surface of ice is rather unusual, and so you might anticipate that friction forces acting upon a slider moving over ice are also unusual. We now apply our knowledge of friction to the problem at hand: determining the friction forces that act upon winter sports athletes.

ICE FRICTION

When a skate or a bobsled runner slides over ice, what type of friction force acts upon it—sliding friction or lubricated friction? Experiments show that the coefficient of friction for steel on ice is in the region $\mu = 0.003-0.007$ for ice skates and $\mu = 0.01-0.05$ for bobsled runners. This is extremely low for solid sliding over solid, which usually produces sliding friction coefficients around $\mu = 0.4$, as we saw earlier. We also saw that lubrication lowers the friction force, so perhaps lubricated friction is operating here. But experiment rules this out: there is little speed dependence for the observed friction force, so this points us back toward sliding friction. Ice is strange.

8. Aerodynamic drag is a subject about which much—too much—has been written. A lot of the popular accounts of drag, and of the closely related phenomenon of aerodynamic lift, are misleading. In my book about the physics of sailing (which involves much aerodynamics) I needed to dispel this misinformation and so included an appendix dedicated to this end. See Denny (2009, 207–31).

Scientists used to think that the phenomenon of regelation provided the explanation for the observed low coefficient of friction. An ice skater generates quite a lot of pressure on the ice, because of the small area of the skate blades, and so it was thought that this pressure would cause the ice underneath a blade to melt, providing a lubricating layer of water. Then, once the skate passed over a section of ice, the pressure would diminish and the ice would refreeze (this is regelation, you may recall). However we now know that this is not the case. Calculations show that the amount of meltwater generated by the pressure increase of an ice skate is negligible— far too small to provide a lubricating layer. We still think that there *is* a layer of water generated by a passing skater or bobsled, but the mechanism is not regelation.

Friction dissipates energy, and much of this dissipated energy is ab-sorbed by the ice as heat. You know that this is the case from every-day experience: vigorously rubbing your hands together in cold weather warms them up (try it now, if you need convincing). Similarly, this *frictional heating* mechanism melts ice beneath a moving skate or runner, producing enough meltwater to provide a lubricating layer. It must be a partial layer at most, and not true lubrication such as is provided by oil in a car engine, because the observed friction force is not strongly dependent upon slider speed. Perhaps only some of the ice melts beneath a skate or a runner (near the back of a skate, where it has been heated for longer than near the front), so that the observed friction force is a mixture of sliding and lubricated friction. Careful measurements of ice skate temperature changes during skating, of the ice track immediately behind a skater, and of the *ejecta* (water sprayed out from a moving skate) support the fric-tional heating viewpoint. On the other hand we know that ice is slippery even if we stand still on it, and this cannot be explained by frictional heating (which requires slider movement). We are forced to conclude that ice friction is not yet fully understood. However, the frictional heating theory is getting us close to the truth, and in practice we know how to deal with ice friction in mathematical calculations. We may effectively treat ice-steel friction as a type of sliding friction with a very low friction coefficient.

More detailed experiments show that the friction coefficient increases as ice temperature falls. (It is a minimum at a temperature of $-7°C$.) So when we calculate the movement of a bobsled down a track, for example, we must carefully choose the friction coefficient to suit the meteorological conditions that apply for the case we are considering. Other experiments

MAKING ICE

No, not ice cubes for your three fingers of whiskey, dispensed from your fridge. Here I mean making ice surfaces upon which indoor winter sports events can take place. Not only is the scale of ice making very different from that of your fridge,* but the method is much more intricate. The final surface is between ¾ inch and 1 inch thick (any thinner and the ice may shatter under impact; any thicker and the top surface may become soft), and it is laid down in 8–10 individual layers, not all at once.

The first three layers are very thin; each is allowed to freeze before the next is added. After the third layer has frozen, the total thickness of ice is only about ⅛ inch. The surface is then painted,† and the remaining layers are added, to bring the thickness up to the final value. For curling, a final layer of water droplets is spread over the top surface to provide "pebbling"—of which, more in chapter 4. During a hockey game the ice takes a pounding, and so the surface is renewed between periods, to fill in cuts.

Ice temperature varies with the sport to be played upon it. Hockey requires hard ice, and the ice temperature in a hockey rink is typically −4°C. Figure skaters prefer slightly softer ice (at −3°C) for better grip and to reduce the likelihood of ice shattering after a jump. Surfaces that are colder than −5°C chip too easily—so making the ice (and maintaining it) at the right temperature throughout the sporting event is critical. It is also tricky: air temperature and humidity within the rink may change significantly throughout the event, as doors open and close, and as thousands of fans enter the arena—bringing in body heat as well as enthusiastic support for their team.

---

*It takes between 10,000 and 15,000 gallons of water to make the surface for an ice hockey arena—that's over two million ice cubes.

†The background color is typically white, to aid visibility, but on top of this there may be additional markings—for example, if hockey or curling is to be played on the ice.

---

show that at temperatures near 0°C (32°F) the friction coefficient for ice skates is $\mu = 0.0046$ for movement on a straight section of track and $\mu = 0.0059$ on a bend. The friction coefficient increases slightly with increasing skate speed, which shows us that lubricated friction is indeed playing a role. However, for the spread of speeds used in speed skating we can effectively treat the friction coefficient as a constant (or as two constants—one for the straights and one for the bends).[9]

SNOW FRICTION

We have seen already that to a physicist, snow is ice with complications. The friction of polyethylene skis on snow is correspondingly more complex than that of the same material sliding over ice. Again, it will turn out to be the case that we can describe snow friction reasonably well as sliding friction, but with a reduced friction coefficient to allow for the unusual physics. Again, frictional melting will play a role.

The friction coefficient for skis on snow has been measured to be $\mu = 0.05$–0.20. This value is lower than we would expect for two "normal" solids but is higher than we found for ice. We can see qualitatively why snow friction lies between ice and "normal" friction. First, skis do not quite slide through snow; to a greater or lesser extent they snowplow. That is, the skis compact the snow beneath them and plow their way through the snowpack. This compaction process saps energy from the moving skis and so appears as a friction force. This extra source of friction makes snow less slippery than ice. Another reason for the increased friction coefficient of snow is the larger area of ski surface, compared with the surface area of skates or of sled runners. "Aha!" you say, "but you told me earlier that sliding friction does not depend upon surface area!" Indeed I did, but I also said that this observation applied only when both surfaces (that of the slider and of the, ahem, slidee) were hard. With snow we have one surface that gives considerably, and the snow deformation means that the effective contact area is very large and increases as the ski area increases. Thus, compaction increases snow friction.

Yet friction is still less than normal because frictional heating plays a role. Indeed, it plays a more important role in snow friction than it did for ice friction. Here's why. First, skis are much longer than skates, and so there is more time for frictional heating to dump heat into the snow and create meltwater. Careful experiments show, for example, that the temperature of cross-country skis oscillates at the skating frequency. That is,

9. For a refutation of the pressure melting mechanism, see Colbeck (1995). For measurements of ice skate friction coefficient, see de Koning, de Groot, and Schenua (1992). The temperature measurements, and details of the frictional heating mechanism, may be found in Colbeck, Najarian, and Smith (1997); Evans et al. (1976); and Mills (2008).

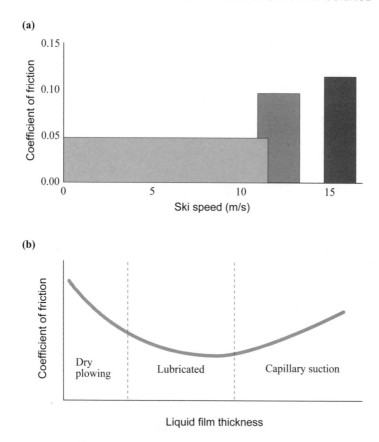

Figure 1.8. Snow friction. (a) Measured friction coefficient for three different ranges of speed, for downhill skiing (18° incline). These data demonstrate that lubricated friction is significant: the friction coefficient increases as speed increases. Data from Nachbauer, Kaps, and Mössner (1992). (b) Schematic of friction coefficient vs. water film thickness, showing that lubrication reduces friction but that too much water causes it to increase (probably because of capillary suction). See Colbeck (1996).

temperature rises when a ski is sliding over snow and falls when the ski is lifted off the snow surface. Also, the temperature of the ski has been observed to increase from the tip of the ski to the tail. Both these observations support the frictional heating idea. As with ice, lubrication by meltwater generally lowers the friction coefficient, though a very large amount of water (i.e., a thick film of water between snow and ski) has been observed to *increase* friction. The minimum friction seems to be for an average film thickness of 8 microns (about 0.0003 inch). Thicker films lead to increased contact area and capillary action that creates suction, holding down the ski and increasing sliding friction. The dependence of the sliding friction coefficient (of skis upon snow) upon ski speed and lubrication film thickness is illustrated in figure 1.8.

But there is more to snow friction. For example, the *color* of skis probably influences the friction coefficient. What? How can this be? Well, some colors absorb more heat from the sun than do other colors. Absorbed heat increases frictional melting. This does not happen on ice because ice skates and sled runners present less of an area to the sun and are usually shaded anyway. Also, skis are thinner. Put all these factors together and it is easy to see why solar heating influences ski friction on snow much more than skate or runner friction on ice.[10]

Snow type affects friction (recall that there are seven types of snow). Perhaps that is not surprising, but how about this: sliding skis become electrically charged, and the amount and distribution of static charge influence friction. Potential differences fluctuate by a volt or so during a ski stride cycle. Charge buildup is not well understood, but it is easily measured. Static charging may attract grit particles that exacerbate friction (and greatly complicate the physics of friction).[11]

10. See Colbeck and Perovich (2004) for the influence of ski color. The effects of solar heating upon ski friction are important in another way. Skiers can feel a difference between the frictional force of a section of snowpack that is in the shade and another section—even one very close by—that is in the sun. Here, it is the snow surface that is being heated, and not the ski. Whichever is heated, the meltwater increases and changes the friction coefficient.

11. Snow friction is being actively studied, in part because it is only incompletely understood and in part because of its importance in understanding how avalanches work. See Colbeck (1996). For measurements of friction coefficient between ski and snow, see Miller et al. (2006); Nachbauer, Kaps, and Mössner (1992); Stefanyshyn and Nigg (2000); and Zatsiorsky (2000).

PART I

Ice Sports

2 SKATING ON THIN ICE

We have seen how thin the ice is—one inch or less—for indoor ice sports such as speed skating and hockey. In this chapter we examine the physics of movement over ice, using ice skates. The act of skating is in many ways unnatural, and yet figure skaters make it look graceful, elegant, and artistic; speed skaters are the fastest people on two feet; and hockey players create a frenetic mayhem on ice in the fastest-paced team sport in the world.

THE SKATES

If you are not a skater, you would be forgiven for thinking that an ice skate is simply a boot with a thin metal blade attached to the sole. In fact, ice skate design is much more complicated than that, as ice skates have evolved specific forms to match different conditions and requirements—a testament to the long history of ice sports. Before describing these forms, there are a couple of general features that I need to tell you about.

In figure 2.1a I have sketched an ice skate blade as seen from the side and from the front. From the side, you can see that the blade is not straight— the section that is in contact with the ice is curved upward at both ends. The blade is known as the *rocker* and its radius of curvature, which is different for different sports, as the *rocker radius*. This radius can change along the blade: for some types of skate it is less at the front of the blade than at the back. From the blade end you can see that the bottom surface is not flat, but instead is hollowed out. The depth of the hollow is described by the *radius of hollow* which, like the rocker radius, changes from sport to sport. The blade material is made of carbon steel, which is very hard.

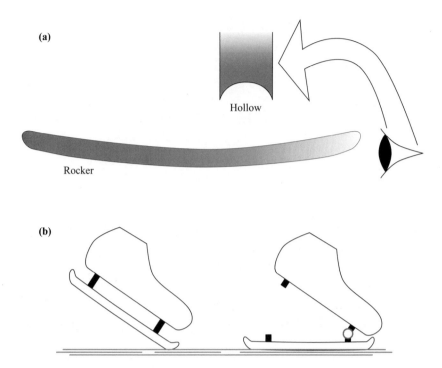

Figure 2.1. Ice skate blade design. (a) When viewed from the side, the blade is curved, with a rocker radius that varies from sport to sport. Viewed from the front, the blade is hollowed out, to provide sharp edges for gripping the ice. The radius of hollow varies with the sport and the skater. (b) During the power phase of a skating stride, the hinged blade of a clap skate (*right*) stays in contact with the ice for longer than the rigid blade of a traditional skate (*left*).

Despite being so hard, ice blades frequently require sharpening, to maintain the edges on either side of the radius of hollow.[1]

What is the purpose of these two types of curvature in a skate blade? A very curved rocker makes it easier for the skater to turn, and so sports that require great maneuverability, such as hockey or figure skating, are played on short-bladed skates with a short rocker radius. There is no free lunch: reduced stability is the price paid for this increased ease of turning. These characteristics make sense; there is greater forward and backward freedom of movement for blades with short rocker radius than for flatter blades with a long rocker radius, accounting for maneuverability with the former

1. An interesting article on the importance of skate sharpening can be found in *New York Times* (2009).

and stability with the latter. The radius of hollow gives blades sharp edges and so permits the skater to lean over without slipping, as when taking a bend. Consequently, sports which require the skater to make sharp turns, such as figure skating or hockey, require sharp-edged skates (with a small radius of hollow). Generally, the best radius of hollow for a given skater depends not only upon the sport but also upon the ice temperature and the skater's weight and skill level—less skilled skaters are better off with large radius-of-hollow skates.

Now that you know about the general features of ice skates, I can describe to you the different skate designs that have evolved for different ice sports. Figure skates (fig. 2.2) are very sturdy, with a blade that is quite thick, at 4 mm, and with strong ankle support. The rocker radius is small, at 2 m, and is even smaller near the front of the blade, to facilitate spins and turns. Figure skate blades are short, and their most striking feature, evident in figure 2.2, is the serrated *toe pick* at the front, which digs into the ice and so provides stability during the launch phase of a jump.

Hockey skates have blades that are about 3 mm wide, with a rounded heel. The boot is a reinforced shell that supports the skater's ankles and protects against cutting from the skates of other players. The rocker radius

Figure 2.2. Figure skates. Note the serrated toe picks.

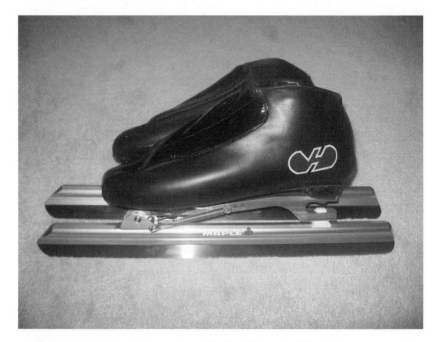

Figure 2.3. Speed skates feature long blades. These are clap skates.

varies over the range 3–4 m, being shorter for forwards, who have to maneuver a lot, and with a *balance point* further toward the front for defensemen, who need to skate backward a lot.[2] Goalie skates are lower to the ice, to prevent the puck from sneaking underneath.

Speed skates are very long (fig. 2.3). Nordic touring skates are extremely long—50 cm (20 inches)—to provide stability on uneven natural ice. The width or thickness of speed skate blades is much less than the blades of figure skaters or hockey players—as thin as 1 mm—to minimize friction. Long-track speed skates have a big rocker radius, at 22 m, but short-track speed skates have a much smaller 8-m rocker radius, due to the tighter bends. For the same reason, short-track skates are higher off the ice, so that the skater can lean over more. Speed skates have an infinite radius of hollow—in other words, they are flat at the bottom. This feature reduces the sliding friction coefficient, but makes for more difficult maneuvering.

Clap skates are hinged at the front, as we saw in figure 2.1b. The speed skates shown in figure 2.3 are clap skates. This Dutch innovation of the

2. The balance point of a skate is the rocker point of contact with ice at which the skater most comfortably stands upright.

1990s has revolutionized speed skating: clap skates take half a second off the lap time (of a 500-m lap) during speed skating events, and so world records tumbled when these skates first appeared on the international scene.[3] The name comes from the clapping sound made when the spring, acting at the hinge, causes the blade to clap against the sole of the boot when raised off the ice. Clap skates increase the time of contact of skate and ice during a power stroke, without increasing the drag caused by digging the toe end into the ice.

I will explain how speed skaters power their way around a course soon enough; here, I will mention only the odd and rather unnatural movements that are required, which explain why clap skates are better for speed skating than are the older rigid skates. When we accelerate from a standstill to a run over firm ground, we naturally lean forward and push off the ground with the front of our feet. On ice this does not work very well because the toe end of our skates would drag against the ice and retard forward movement. So skaters learn to push off the ice with their heels. Certain muscle movements—giving rise to the powerful ankle extension, natural when running—have to be suppressed when skating with rigid skates. Clap skates permit a more natural running action when accelerating over ice.[4]

SPEED SKATING TECHNIQUE

We have looked at the skates that athletes on ice wear. Now let's analyze the motion of a speed skater when starting up, moving along a straight line, and moving around those tight bends.

The Push-Off

To bring you up to speed, so to speak, I need to start you off: to show how to accelerate on ice from a standing start. Speed skaters do this, and they get up to high speeds very quickly, beginning with a frantic-looking run and then progressing into a smoother gliding motion, before reaching a steady speed. The initial run looks so frantic—like a drunken giraffe on roller skates—because the actions involved are so very different from the

3. Clap skates were actually developed in the 1980s in Holland, but they became commonly used in international competitions only during the 1990s.
4. See de Koning et al. (1989, 1995).

natural actions of running: the knees are bent, to provide a low center of gravity; the feet push out sideways more than backward, and skaters are taught to pick up their toes before their heels, to avoid toe drag; the feet are brought close together prior to the power stroke. Of course, the reason the skating action is so different from running is because of the slipperiness of ice. If you tried to run on ice from a standing start, you would begin by pushing backward. You are relying on Newton's third law: the ground would push back at you with equal force, propelling you forward.[5] On ice, however, there is not enough friction to provide a grip, so the ice cannot push you forward; instead, you lose your footing.

Ice skates are designed for low friction when the skater is gliding forward (so that trying to run on skates is very difficult indeed). They are also designed so that the resistance to slipping sideways is very great. Combine these two design features, and you see that the best way to make forward progress is by moving sideways, as shown in figure 2.4a. Here, we have a skater whose right skate is moving along the ice in a straight line that is at 45° to his intended direction of motion. His next power stroke begins when he brings his left foot close to the right and then pushes off in a perpendicular direction, as shown.[6] There is no slipping of the back foot, because the force F^* is sideways to the blade. Now Newton can join the fun by providing a reaction force, the force F (equal in magnitude to F^*, but in the opposite direction), so that when the skater transfers his weight to the left foot, he glides in the direction shown. In this way, the skater makes forward progress via a series of steps that are each at 45° to his intended direction of motion. So, you can see that if our skater has a skate speed v, his body speed along the intended direction of motion is $v \cos 45 = 0.7v$.

So this is how skaters progress along the ice. Perhaps surprisingly, the best angle for the skate to make with the intended velocity direction is not 45°. In technical note 2, I show that the most efficient use of skater power results if the angle a of figure 2.4b is 35°. This angle results in about 77% of

5. Newton's third law is usually paraphrased as "for every action there is an equal and opposite reaction." I have assumed that readers who wish to follow the technical notes are familiar with Newton's three laws; if you would like to learn more about them, a gentle introduction in the context of sport can be found in Goff (2010).

6. Our skater happens to be male. I will ensure gender equality by considering our skier, in a later chapter, to be female. Unfortunately, the ski jumper must be male, at least for the Winter Olympic Games, because in these games, at the time of writing, no female jumpers are permitted.

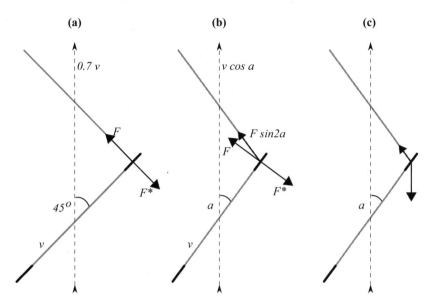

Figure 2.4. Accelerating over ice. Gray lines show skate tracks; black lines indicate the position of the blades at the beginning and end of a stride; dashed lines are the skater's direction of motion; v is skate blade speed over the ice. (a) If the skate moves at 45° to the intended direction of motion, the skater's push force, F^*, will not cause the blade to slip because the force is perpendicular to the blade. (b) Skating at a different angle, a. The component of force along the blade direction is $F \sin 2a$. The best choice for a is about 35°. (c) Trying to skate the way you run, by pushing backward, reduces efficiency.

the power expended being used to propel the skater in the desired direction, compared with about 70% when $a = 45°$. One immediate consequence of this optimum angle is the requirement that the push angle, the angle at which the skater directs his push force F^*, be more nearly sideways than backward. What if the skater tries to improve his action by pushing backward, as shown in figure 2.4c? Naively you may suppose that this would be a good idea because the reaction force is then in the forward direction, which is where you want to go. This idea works for running but not for skating. In technical note 2 we find that the efficiency when pushing backward is a maximum of 56%, and this occurs when the angle a is 52°, which would lead to a Charlie Chaplin–style waddle in the forward direction. Compare this with the sideways push of figure 2.4b, which, we have already discovered, is more efficient; and with $a = 35°$ this optimum efficiency applies at a higher speed, in fact at 82% of the skate speed.

Figure 2.5. An accelerating speed skater. The arms are thrown out wide because the stride is partly sideways. Note the crouched position, to minimize drag. I am grateful to Adnan Hussain for permission to reproduce this image.

After the frantic initial rush, the skater continues to accelerate for some distance (it takes over 100 m to get up to top speed) but now adds gliding and big arm swings to the powerful push, as seen in figure 2.5. Once top speed has been reached, then the skater exerts less energy per stride because he needs only to overcome the effects of sliding friction and drag, and no longer needs to increase speed with each stride.[7]

7. For technical accounts of the acceleration phase at the start of a speed skating race, see, for example, Jobse et al. (1992) or de Koning (1989, 1995).

Maintaining Speed in a Straightaway

Now we are in a position to investigate how a skater applies his energy to accelerate and to maintain a desired speed. Here, we will assume that he is skating along a straight section of track so that we do not need to consider the complications that arise when he hits a bend.

In figure 2.6 you can see two illustrations of the tracks left by a skater, assuming he makes short strides (fig. 2.6a) and long strides (fig. 2.6b). Both of these strides occur in racing: the short strides are used when the skater is accelerating or when he is maintaining a very high speed. Longer strides are used in longer races, when he needs to conserve energy. In figure 2.6c you can see a graph of short-stride skating speed and long-stride skating speed. These plots were made assuming the skater speeds calculated in technical note 3, for both the power phase of a skater's stride, and the gliding, or coasting, phase. Clearly, the shorter stride has a higher average speed.

Before I show you why different stride lengths apply in different types of races (shorter average stride length for short-track speed skating and longer strides for long-track events), we need to appreciate a fundamental fact about the nature of skating (or of skiing or running—indeed of any human locomotion over the ground). There is a maximum speed that a skater can attain, which depends upon his endurance, strength, and skating technique, as well as upon the sliding friction and drag. The point I need to emphasize is that the *existence* of an upper speed limit is fundamental. Even if we assume that the skater is extremely strong and of limitless endurance, with perfect skating technique, and assume zero ice friction or aerodynamic drag, there would still be an upper limit. This upper limit is determined not by the finite strength or endurance of the athlete, but by the manner in which a skater (or runner) accelerates. He pushes off the ice with a certain foot speed, which is physiologically limited to about 43 kph: he cannot push faster than this.[8] As he moves down the ice, picking up speed, he is less able to push hard against the ice. When his speed over the ice reaches the speed at which his leg can push, the net

---

8. In the text I provide speeds in kilometers per hour (1 kph = 0.62 mph); this is the unit of speed adopted by Olympic sports. In graphs and in the technical notes I use the standard metric unit of speed, meters per second (ms^{-1}).

(a)

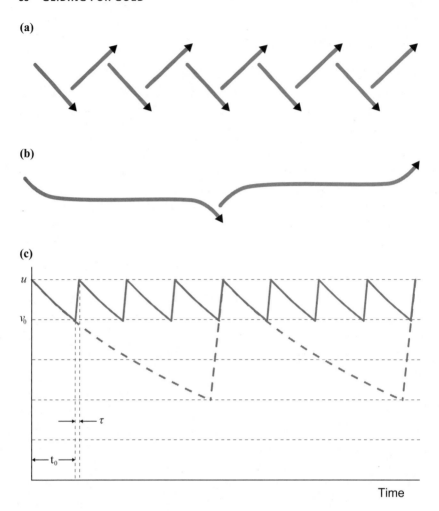

(b)

(c)

Figure 2.6. Skate strides. (a) When the skater is accelerating, short, rapid strides are the most efficient. (b) Over long distances, longer gliding strides save energy. (c) Speed vs. time for boost-coast short and long strides. The long stride is represented by the dashed line.

result is that his leg is exerting no force on the ice. So the skater cannot accelerate any more, and his top speed has been reached.[9]

Given that there is a top speed for every skater, consider again figure 2.6c. Clearly the average speed for a short stride is closer to this top speed (denoted u in fig. 2.6c) than is the average speed for a long stride. So why

9. For the maximum possible human leg speed, see Haché (2002).

TAKE IT TO THE MAX

Now that your mind is finely honed in the subject of speed skating, through a rigorous but rewarding training program (in other words, by reading this book), you may have asked yourself a question: "If the maximum human foot speed is 43 kph, then how come speed skaters can reach speeds of 50 kph?" Good question. The answer is that the maximum foot speed is relative to the other foot. When a runner plants one foot firmly on the ground, he can move the other foot relative to it (and so relative to the ground) at no more than 43 kph, which provides an upper limit for human running speed. *But* skating is not running, as I have tried to emphasize. In particular, the skater's foot on the ice is not planted firmly. Consider figure 2.4a once again. The "planted" ice skate is moving at speed v over the ice, at an angle of 45° to the skater's intended direction of motion, which means that the "planted" skate is moving forward at a speed of about $0.7v$. The next stride also contributes $0.7v$ to the skater's speed, and so the total speed is $1.4v$. This is why maximum skate speed can exceed maximum leg speed: if $v = 43$ kph then the theoretical maximum skating speed is about 61 kph.

do skaters bother with long strides? Long strides save energy and allow the skater to regain his breath. Thus, in long-track speed skating we see a skater push off with one leg against the ice, and then glide, or coast, for some considerable distance before pushing off with the other leg. The stride looks leisurely and pedestrian, especially as the skater looks quite lackadaisical, with his hands behind his back as if he is just casually strolling through the park. In fact, his movement is far from lackadaisical. He leans forward and places his hands behind him to minimize aerodynamic drag. His pace looks pedestrian on TV when viewed from the front, but stand by the trackside and you will see that he is *flying* along. Skaters are the fastest people on earth, if you limit consideration to those athletes who are powered solely by their own muscles (thus excluding gravity or the mechanical advantage gained by riding a bicycle). The 500-m world record speed skater is, at the time of writing, Jeremy Wotherspoon of Canada, who on November 9, 2007, raced around a 500-m track in 34.03 seconds. His average speed was 52.88 kph, which is 45% faster than the fastest 100-m

Figure 2.7. No, not a six-legged skater. Two skaters follow closely behind the leader to benefit from drafting. Thanks to Adnan Hussain for this picture.

sprinter—and Wotherspoon covered five times that distance around a track that included tight bends.[10]

We have seen the main reason why skating is faster than running: a skater's blade is not stationary on the ice when he pushes off for his next stride. There are other, secondary reasons. First, the techniques required for running and skating are very different. A skater may swing his arms when accelerating, as a runner does, but during the glide he saves energy by not swinging his arms. A runner does not have this option. Second, the skater will reduce his aerodynamic drag by leaning forward in a crouch while skating—again something a runner cannot do. Assuming realistic

10. The difference in speed is even more obvious over long distances. At the 2010 Winter Olympics in Vancouver, Dutchman Sven Kramer won the men's 5,000-m long-track speed skating event in an Olympic record time of 6 minutes 14.60 seconds. Two years earlier, at the Summer Olympics in Beijing, Kenenisa Bekele of Ethiopia set a new Olympic record for the men's 5,000-m track and field event with a run of 12 minutes 57.82 seconds, more than twice Kramer's time.

parameters, we can show that a skater who is maintaining speed—not accelerating—is expending something like 80% of his energy overcoming aerodynamic drag and only 20% overcoming sliding friction. (These numbers neglect energy spent internally—swinging arms or breathing, for example; I am considering only the mechanical energy expended to maintain speed.) Third, skaters wear smooth, all-body suits that make them look like Spiderman because these suits reduce aerodynamic drag.[11] Fourth, skaters in a race will follow close behind an opponent to reduce drag, saving energy by slip-streaming (or *drafting*, as it is known in skating); see figure 2.7. In these ways skaters have significantly reduced aerodynamic drag compared with runners, and so they are able to move faster.

Let us return now to the question of stride length. In technical note 3, I show that if a skater wants to minimize the energy he expends during each stride, then he should adopt long strides. In other words, the energy expended per unit distance covered is less for long strides than for short strides. Given that long-distance skating events are tests of endurance, we would expect long-distance skaters to minimize their energy expenditure in this way and adopt long strides. Lo and behold, that is what we see on the ice. On the other hand, short-track skaters will optimize for maximum speed, not minimum energy expenditure, and go for shorter strides. The results of the calculations of technical note 3 are plotted in figure 2.8, where you can see how energy per unit distance, and energy per unit speed, depend upon the duration of the gliding or coasting phase of a stride (the interval t_0 of fig. 2.6c).[12]

Taking the Bends

Having investigated how a skater moves and how he strides along a straightaway, we must now permit him to move along a curved track, because the real-world long- and short-track speed skaters must do so. Indeed, skating the bends properly is an important aspect of the sport. The bends on a short track are tight, with a radius on the inside lane that is as small as 8 m. For long tracks, the bend radius is three times bigger. You may wonder at the fact that the rocker radius on short- and long-track skates is about the

11. The manufacturers of speed skaters' suits keep design details—even the material—a closely guarded secret.

12. To learn more about the biomechanics of skating, see Albert and Koning (2007) or Zatsiorsky (2000).

FEELING THE DRAFT

Data from the 2010 Vancouver Winter Olympics in long-track speed skating events shows us just how important aerodynamic drag is to the sport, and in particular how much speed skaters can benefit from *drafting*, which is to say from slipstreaming behind another skater.

In part a of the figure here, I have plotted the average speed for all the long-track events (from 500 m to 10,000 m for the men's events, and from 500 m to 5,000 m for the women's events).* The data points represent the average times of gold, silver, and bronze medal winners. Three trends emerge from this graph. First, the men skate faster—unsurprising given their greater strength. Second, the average speed decreases with increasing distance. Again, this is unsurprising, given that athletes will tire more in

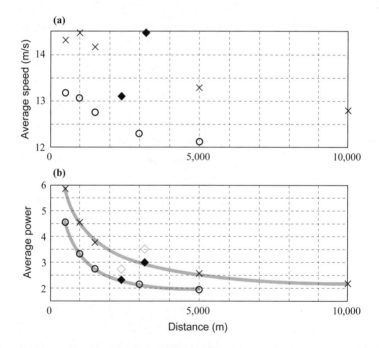

Speed skating statistics for the 2010 Winter Olympics medal winners. (a) Average speed at given distances for men (x) and women (o). The team pursuit events are shown as black diamonds. (b) Average power vs. distance. The team pursuit results are shown as open diamonds: they fit the general curve (gray line) only when the skaters' aerodynamic drag factor is reduced by one-sixth.

longer-distance races. One exception, note, is the men's 500 m, in which the average speed is less than that for the 1,000 m. This can be attributed to the time that it takes for men to accelerate to their top speed—more time than for women because of their greater mass and higher top speed. Third, the team pursuit events buck the trend: average speeds for these events is greater than for the other long-track speed skating races.

To explain what is going on with team pursuit speeds, I should say something about these kinds of races. In team pursuits, instead of having one skater per lane racing against the clock, we have three skaters per team (with the teams in the same lane but on opposite sides of the oval so that they don't have to pass each other). The three skaters in each team follow close together, one behind the other, to benefit from drafting, and perhaps also to push the teammate in front if he is flagging. Lead changes often, because the lead skater tires more quickly. The winning team is the one whose *last* skater finishes in the shorter time.

In part b of the figure I show the average power that each skater exerted during the Olympic finals. Note the smooth trend of power falling off at longer distances. Again, team pursuit events stand out, if we assume the same aerodynamic drag. To bring the team pursuit data points onto the line, I had to reduce the average aerodynamic drag experienced by team pursuit skaters to 83% of that experienced by single skaters.[†] This shows that drafting reduces the power expended on overcoming speed skaters' aerodynamic drag by one-sixth. The data also show that about three-quarters of the skaters' power was used to overcome drag and only about one-quarter to accelerate body mass (at the beginning or around the bends).

[*]At the Vancouver Games, the U.S. women's team pursuit skaters upset the heavily favored Canadian women by knocking them out of medal contention. Canada gained revenge in the men's final, narrowly beating Team U.S.A into second place. One odd characteristic of the veteran American team pursuit captain, Chad Hedrick, is that he skates with his tongue hanging out, like a hunting dog. However, Hedrick was faster than any dog could be over eight laps (two miles).

[†]My calculation has not been adjusted for the slightly shorter lap distances that occur in team pursuit events, where, unlike other long-track races, skaters do not have to alternate between inner and outer lanes: team pursuit events take place on the inner lane. On the other hand, I have also not adjusted for the fact that the *last* skater, not the first, determines the race time in team pursuit, or for the extra distance these competitors skate when changing position. These partially cancel the reduced lap distance effect.

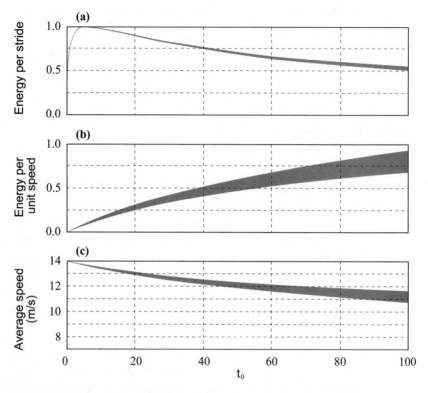

Figure 2.8. Skate stride efficiency. (a) The energy expended per unit distance covered is least for long strides. The time t_0 is the coast-phase duration in figure 2.6c. (b) The energy expended per unit of generated speed is a minimum for short strides. (c) Average speed is higher for short strides, assuming the same maximum skate speed. In these plots, the assumed maximum speed is 14 ms^{-1}; the drag factor, b, is 0.001 m^{-1}; and the q value (a parameter of technical note 3) varies between 5 ms^{-1} and 9 ms^{-1}, corresponding to a friction coefficient of about 0.005.

same as the bend radius on the tracks. This is not a coincidence. If you think about it, when a skater leans into a bend, he wants his skates to describe a curve on the ice of the same radius as the bend to help him turn through the bend with minimum sliding friction.

In addition, short-track events are always arranged so that the skaters move counterclockwise around the track, and their skates are adapted to take only left-hand bends. The position of the blade on the sole of the boot is moved—it is offset from center to the left—so that the skater can lean into those bends further without the side of the boot coming into contact with the ice. Such contact, of course, would increase sliding friction and so slow down the skater. (It would be interesting to see how well a short-track speed skater could negotiate the bends if he were obliged to skate around the track in a clockwise direction.)

We can calculate how much a skater has to lean over when negotiating a bend at speed; see technical note 4 for details. If our skater takes a 8-m-radius short-track bend at a speed of 36 kph, he will have to lean into the bend at an angle of about 50°. He can skate faster around a 25-m long-track bend (say at a speed of 45 kph—typical for a 5,000-m long-track event) by leaning over at about 30°. These are the angles we see in practice.

There are two methods that a skater can use to negotiate a bend. First, he can lean into the bend and skate around on one leg, letting the rocker angle of his skate naturally point him in the right direction. Of course, if he does that, he is gliding over the ice, not powering his way, and so he will lose speed due to drag and friction—but will gain some breathing space. For short-track events, skaters often can be seen skating a bend in this manner.[13] The second method, sometimes used in short-track events and always in long-track events, is to power around the bend using both skates. This is achieved as shown in figure 2.9.

In the second method the skater travels a short distance (denoted l in fig. 2.9) in a straight line on one skate. Near the end of this stroke he pushes toward the center of the circle (approximately perpendicular to the skate blade) and sets his inside skate onto the ice in a slightly different direction, as shown. The bend is part of a circle (of radius R in fig. 2.9), and the inside skate is the one nearest the center of this circle. In this manner the skater powers around the bend via a series of short glides, each in a straight line but with the succession of glides following the curve of the bend. If he pushes toward the circle center with a force F, then a small component of this force (labeled $F \sin a$ in fig. 2.9) is along the direction of his next stroke. So the skater can maintain speed if this component of his force matches the sliding friction and aerodynamic drag that act upon him. In technical note 4 I do the math for you and show that the skater must exert a force that is perhaps three times as great as the force he must

13. For short track, recall, only the skater's finishing place matters, not the time, so that a skater in the lead may well choose to get around a bend by gliding on one foot. If his opponents want to pass him, they will have to take a longer route further out from the center of the bend. Often, in fact, a trailing skater will opt to glide through the bend, if the leader does so, and will then seek to gain the lead on a straightaway section of track.

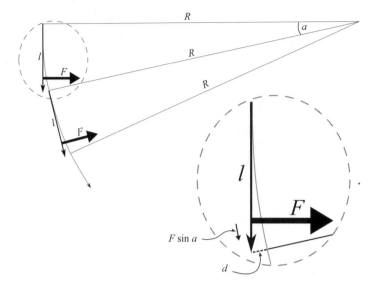

Figure 2.9. Skating a tight bend. The skater pushes perpendicularly to his skate blade—almost toward the center of the bend circle. Each stride covers a distance l; R is the circle radius; F is the push force. *Inset*: with each stride, the skater steps a distance d toward the circle center.

exert on the straightaway to overcome these dissipative forces and maintain his speed.

The speed with which a skater can fly around a bend, and the tightness of the bend he can take by powering his way, is limited by his size. We can understand why this is the case by considering the way in which he steps from one stride to the next. You can see from figure 2.9 that he must step a distance d, toward the circle center, as he switches from the right to the left skate, and again when switching from left to right. Bends are always taken counterclockwise, and so the skater's left leg is always on the inside. This makes it easy when he switches from right skate on the ice to left skate on the ice. He is leaning leftward, into the bend, and so it is easy for him to put his left foot down nearer the circle center: it must feel like he is climbing uphill.

However, when he is switching from the left to the right skate, things are more difficult because he must cross his right leg over his left as shown in figure 2.10 in order to plant the skate on the inside, closer to the circle center. In technical note 4 I have assumed that the distance d is the same for both cases, but in reality it probably isn't, so my d is an average value.

Figure 2.10. Skating a bend. The right leg must cross over the left, as all bends are taken counterclockwise. Thanks once again to Adnan Hussain.

The length of stride that a skater takes, and thus the number of strides he must make to go around the bend, depend upon how large a distance d he can step, which depends upon his leg length. Counting the number of steps that skaters take to get around a bend leads to the conclusion that a skater steps an average distance toward the circle center of $d = 0.4$–0.5 m (see technical note 4 for the derivation). So, just as a runner's speed is limited by his leg speed and stride length, so the speed with which a skater can negotiate a bend is limited by the same factors, though the required movements are very different.

FIGURE SKATING MANEUVERS

Figure skaters do it with style. They are not about speed or time; they are all about presentation. Think ballet dancers on ice, but don't for a minute think that these performers are just dancers or artists—they are athletes.

In figure skating you will see a bewildering succession of turns and jumps and spins, with an equally bewildering technical vocabulary to describe it all. For example, there are eight basic ways in which a skater can glide down the ice in a straight line: on the left or right foot, moving backwards or forwards, sliding on the inside or outside edge of the skate blade. So, for example, a figure skater's glide may be described as *LBO*, meaning *L*eft foot, *B*ackward, *O*utside. Our skater may then, for example, perform the two-foot turn described as an *LBO-RFI Choctaw*.[14]

There are one-footed turns as well, performed on the front part of the skate blade, which has a smaller rocker radius than other parts, facilitating such turns. One such turn is named a *three*, because the shape of the trace left on the ice is like a figure 3.[15] Other distinct one-footed turns include the *counter*, the *rocker*, and the *bracket*, which traces a line like a ty-pographical brace, {.

A figure skater performs a large variety of different jumps. Like the turns, jumps can be cut and diced in different ways for classification purposes. Thus, there are *edge jumps* (such as the *Axel*, the *loop*, or the *Salchow*), in which the jump is assisted by the edge of the skate on the free foot (i.e., the foot that is not providing the main power for the take-off), and *toe jumps* (such as the *toe loop*, the *flip* and the *Lutz*), in which the jump is assisted by the toe pick of the free foot. Jumps mostly involve rotations and are described as either *natural* or *counter-rotated*. In a natural jump, the rotation is in the same sense as the curve on the ice immediately before the jump. In the hieroglyphs of figure skating's technical vocabulary, the Axel jump is listed as *fo 1½ Tbo*, meaning that it is launched in the forward direction from an outside edge and that the skater then makes 1½ aerial revolutions in the natural direction before landing backward on an outside edge.[16] An expert skater can perform triple jumps, in which he or she makes three complete rotations while airborne. Some men can perform quad jumps—four rotations. Quad jumps require being airborne for a long time and hence require great leg strength for the launch.

14. For the *Choctaw* the skater changes feet, direction, and blade edge, whereas in the other well-known figure skating two-foot turn, the *Mohawk*, he changes feet and direction, but maintains the same edge. The label "Mohawk" dates from the nineteenth century, for a fanciful resemblance by the trace left on the ice to an Indian bow. The Choctaw was subsequently named as a variant of the Mohawk. For a history of figure skating, see Copley-Graves (1992).

15. The name "figure skating" itself comes from the practice of skating around the ice in such a way as to leave traces of the figure 8.

16. A more complete description of figure skating turns, jumps, and spins can be found on the comprehensive figure skating Web site, www.sk8stuff.com, which includes video footage of the different maneuvers.

QUAD JUMP

Only a few male figure skaters can successfully execute the spectacular quad jump. The skater must launch himself high enough into the air to be able to make four complete rotations before landing. This element earns *big* points but is risky: in the 2010 Olympics many skaters fell trying to execute this jump, and some balked, reducing a jump from a quad to a triple. The figure shows how much height a skater must achieve to execute a quad or a triple; it also depends upon how fast he spins. In practice, the fastest spins seem to be about five rotations per second (5 Hz). The calculation on which the graph is based is presented in technical note 5.

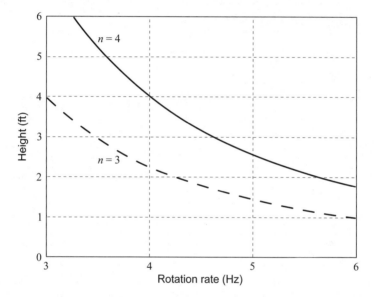

Figure-skating quad and triple jumps. The required jump height is plotted against skater rotation rate for a quad ($n = 4$) and a triple ($n = 3$) jump. The maximum realistic rotation rate is about 5 Hz.

Figure 2.11. Figure skating singles elements. (a) Inside edge spiral. (b) Catch-foot layback spin. Note that the point of contact of the skate blade with the ice is near the front, where the rocker radius is small. Thanks to Caroline Paré for permission to reproduce these images.

Jump rotations can be either clockwise or counterclockwise. The choice is up to the skater; the judges of figure skating competitions have no preference one way or the other. Curiously, most skaters prefer to rotate counterclockwise; the 10% who opt for clockwise rotations tend to be left-handed. Go figure.

Most skaters spin in the same direction as they jump—again, there is no preference in judging. Spins are classified according to the foot used and the skater's normal direction of spin. Thus, for a counterclockwise spinner, a *forward* spin takes place on the left foot; the right foot is used for the more difficult *backward* spin. The skater's body is in one of three positions during a spin: upright (as in the *Beillmann* spin), sitting (e.g., the *broken leg* spin), or camel (e.g., the *flying camel*). For a *camel* spin the skater's body and one leg are horizontal, so that the skater is T-shaped. During spins, the ice skate blade that supports the skater (the low-rocker-radius part, near the front) describes small circles on the ice, so that spins are not true rotations about a point; the shape of blades make true rotations impossible. Two singles elements, including a spin, are shown in figure 2.11.

These three basic types of figure skating maneuver—turns, jumps, and spins—must of course be connected in a sensible and appealing manner. There are many so-called *connecting elements* that achieve the transition from one maneuver to another (*shoot the duck, Russian split*—I leave it to the interested reader to investigate these intriguingly named connecting elements further). In addition, there are elements of figure skating that are unique to pairs events. Thus, for example, there are over a dozen different types of lifts and several variants of the *death spiral*. Two elements from pairs figure skating are shown in figure 2.12.

FIGURE SKATING PHYSICS: ANGULAR MOMENTUM

Physics professors find that the spinning elements of figure skaters' choreography are very useful demonstrations of angular momentum and frequently invoke the example of spinning skaters to get this key physics concept across to their students. In certain physical interactions that do not involve dissipative forces, some properties are *conserved*. To a physicist, the word "conservation" has a slightly different meaning from that applied by ecologists or environmentalists, who sometimes say "conservation" when they mean "preservation." When a physicist says that energy is conserved in a physical system, she means that the total energy of the system is constant, though the forms of energy within the system may change with time. For example, a swinging pendulum may be considered to be a system in which energy is conserved, if we ignore the effects of friction and aerodynamic drag. Energy changes from kinetic (the energy of

Figure 2.12. Figure skating pairs elements. (a) Back inside death spiral. The toe pick of the man's left skate provides the pivot. (b) I don't know the name of this element, but it looks uncomfortable for the man. Thanks again to Caroline Paré for these pictures.

movement) to potential (stored gravitational energy) as the pendulum bob moves from the bottom of its trajectory to the top. Then the energy moves back from potential to kinetic as the pendulum swings back down again. The point is that, despite the fact that the energy within the pendulum sloshes around between kinetic and potential forms, the sum of both types of energy is a constant.

Many interactions involve the conservation of momentum. For example, if two hockey pucks on ice collide, the total momentum that they carry away from the collision is the same as the total momentum they brought into the collision (again, ignoring dissipative forces). A third such physical quantity is angular momentum. In certain physical systems, such as a skater spinning on ice, angular momentum is—more or less— conserved.

Angular momentum is the tendency of a rotating body to keep rotating. A spinning body has a high angular momentum if it is hard to stop it from spinning. High angular momentum may derive from spinning the body very fast or from spinning a very heavy body. We are all familiar with the concept of angular momentum conservation, in practice if not in theory. We know that a spinning skater can increase her angular speed—the rate at which she turns—by drawing in her arms and legs.[17] You may reproduce this behavior in a modest way by spinning on your office pedestal chair. First, get spinning with your legs extended, and then draw your legs in toward the chair axis. This action will cause your office colleagues to stare and you to spin faster. Extend your legs again and you slow down. If your boss complains, just say that you are learning some physics.

The manner in which an ice skater demonstrates the (approximate) conservation of angular momentum is modeled mathematically in technical note 5. This analysis shows that a skater can easily double her initial spin rate by drawing in her arms. Greater increases will arise by drawing in an extended leg. The initial spin rate depends upon, among other factors, ankle strength. The skater must plant a blade upon the ice and twist sharply, before raising herself onto the low-rocker-radius part of the blade so that she rotates in small circles. Then the skater must draw in all the extended limbs so that no part of her is very far from the axis

17. Here I am departing from my convention (in this chapter) of referring to skaters as male because the record for high spin rate among skaters is held by a female. Natalia Kanounnikov has achieved a spin rate of 308 rpm (a shade over 5 Hz), and this feat has been recorded. You may watch it on YouTube, at www.youtube.com/watch?v=AQLtcEAG9v0.

of rotation. This action will maximize spin speed, due to angular momentum conservation.

In this chapter we have examined the physics that underlies skating on ice. Here is a theme that will crop up more than once in this book. It is a drum I like to bang on again and again: appreciating something about how physics constrains movement enhances the sports enthusiast's appreciation of winter sports. Before I researched the physics of skating, I was unaware of the biomechanics of moving over ice on skates, of just how unnatural and counterintuitive it is.[18] Having taught myself (and you, I hope) how skating works physically, I now marvel more, and more knowledgeably, at the skills of skaters. I have always been impressed by a hockey player's ability to maneuver wildly while steering a puck around defenders, but now I also appreciate the skills and strength required to spin rapidly through the air during a backward jump from the ice (most figure skaters take off backward). I now comprehend the skating skill and stamina that is needed to travel 3 miles in 6½ minutes while doubled up.

18. My personal interaction with ice had previously been confined to sliding tentatively over it during games of curling—and dropping chunks of it into glasses of scotch.

3 DOWN THE SLIPPERY SLOPE

The bobsled (*bobsleigh* in Canada and Europe), the skeleton, and the luge all use the same track, though with different starting points (as is illustrated in the plan of the Whistler Sliding Centre track in British Columbia, Canada, in fig. 3.1). These events differ in the construction and weight of the sled, in the positions adopted by the sled rider(s), in the manner of starting off, and in the techniques for steering and braking.

Children of all ages sled for fun, of course. Most of us can remember being pushed down a snow-covered slope as a kid, on a sled of one type or another, and falling off at high speed. As adults, most of us sled surrogately—by proxy—preferring the comfort of a sofa in front of the TV (which is softer and doesn't move so fast) as we watch highly trained athletes sled for fun and Olympic glory. If I were to be suddenly transported from my cozy sofa to a bobsled that is hurtling down the Cresta Run, I guess that sheer terror and exhilaration would crowd out any thoughts I might have about the science of sledding. Happily (or sadly) I

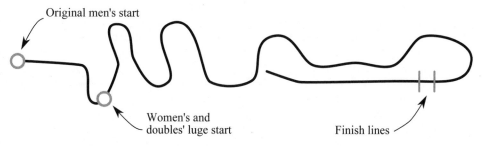

Figure 3.1. Plan of the Whistler Sliding Centre track in British Columbia, Canada, site of the 2010 Winter Olympics. This track is up to 1,450 m (4,800 ft) long with a drop of 152 m (500 ft).

THE HISTORY OF COMPETITIVE SLEDDING

Sleds have been used recreationally, as well as for winter transport, for centuries in both Europe and North America. The *sport* of sledding, however, took off only in the late nineteenth century. Ex-pat Brits living in Switzerland took a liking to tobogganing, as they called it.* The winter resort of Davos gave its name to the first toboggan club, founded by J. A. Symonds in 1883. These well-to-do Englishmen built (or had built) a piste that ran from Davos to Klosters, a distance of 3.2 km (2 miles), and staged a contest involving 21 competitors from seven countries. A rival group of British expats based in nearby St. Moritz, encouraged by local entrepreneurs, constructed a rival track that came to be called the Cresta Run, which hosted its first competition in 1885. The Cresta Run was subsequently used in two Winter Olympics and is still in use today. This run is 1.2 km long (1,100 yd) and drops 157 m (514 ft) from start to finish.†

In the same decade, sled design converged upon the three versions that we know today as bobsled, luge, and skeleton. The bobsledder sat *in* his sled; the luger sat *on* his sled, feet forward; the skeleton sledder lay prone on his sled, head forward. Bobsled was by far the most popular—a fact reflected in its early appearance in the Winter Olympics (beginning with the Chamonix games in 1924).‡ Luge has been an Olympic event since the Innsbruck games of 1964. Skeleton appeared at the St. Moritz games of 1928 and 1948 but has been a regular fixture of the Olympics only since 2002. Since 1923 these sledding events have been governed by the Fédération Internationale de Bobsleigh et de Tobogganning (FIBT).

---

*The word "toboggan" comes from an Amerindian word, whereas "sled" is of Germanic origin. Sleds have runners whereas toboggans are flat-bottomed.

†For more detailed accounts of the origin of sledding events see, for example, the official IOC Web site (IOC 2010) and the St. Moritz Toboggan Club Web site (SMTC 2010). See also the articles on bobsledding, luge, skeleton sledding, and tobogganing in *Encyclopedia Britannica* (Britannica 1998) and *Encarta* (Microsoft 2005).

‡The Chamonix bobsled event was restricted to four-crew sleds. The two-crew bobsled appeared in 1932 at the St. Moritz games. Women's bobsled (two-crew) did not appear until the Salt Lake City games in 2002. Though women had been encouraged to participate in the earliest sledding competitions in Davos and St. Moritz, they were barred from Olympic events from 1924 on. Perhaps this was because of the danger of this sport: overturning toboggans and toboggans that flew off the track had led to a number of serious injuries, and it was decided (by men, of course) that the fairer sex should not be put at risk of such injuries.

have never been so transported, and so am able to ponder the interesting science that describes these high-speed sports. Before beginning my analysis of sledding physics, though, I need to describe the relevant aspects of sled design and of competition rules, and to convey some of the achievements of world-class competitors in all three sledding categories.

BOBSLED, LUGE, AND SKELETON COMPETITIONS

All sleds used in sports competitions consist of steel runners supporting a metal and fiberglass frame. Some modern sleds utilize high-tech composite materials instead of fiberglass. Bobsleds (fig. 3.2) have four runners; luges and skeletons have two each. Bobsled runners are separated by 0.67 m (about 26 in); skeletons and luges are much smaller sleds so that, for example, skeleton runners are between 0.33 m and 0.38 m apart. There are weight constraints that must be satisfied for competition events (table 3.1).[1] Thus, the minimum weight for a two-crew bobsled is 170 kg (374 lb);[2] for a four-man bobsled (there is no four-woman Olympic bobsled event) the minimum sled weight is 210 kg. For competitions with all three types of sleds, if the weight of the sled plus crew is less than the permitted maximum, extra ballast may be added. This is desirable because, as we will see, heavier masses slide faster.

Bobsled and skeleton races start with a run-up. The bobsled crew push the sled as hard as they can for the first 15 m and then leap aboard; the skeleton athlete runs with his sled, then throws it onto the ice and throws himself or herself on top (fig. 3.3). The purpose of the run-up is, of course, to generate speed: a flying start is all-important, as we will see. The luger has no run-up; he sits on his luge and pulls on handles beside the track to propel himself forward, then lies back to reduce aerodynamic drag (fig. 3.4). Bobsleds are equipped with both steering and brake. Luges have no brake; steering is performed by flexing the runners (achieved by using leg muscles) or by shifting one's weight. Skeletons have no steering except that achieved by shifting weight; braking is applied by dragging the feet.

1. For sled dimensions see, for example, Judd (2009).

2. Readers who are scientists will note that I am using metric units of mass (kilograms) rather than of weight. This is conventional. To convert mass to weight, multiply by the acceleration due to gravity, $g = 9.81$ ms^{-2}. The weight of one kilogram is about 2.2 pounds. It is clumsy to provide both mass in kilograms and weight in pounds, so in what follows I have used just metric units.

Figure 3.2. Bobsled action. (a) Push-off phase of a four-man bobsled run at the 2009 World Cup. The push-off bars are retracted when the athletes are inside the sled. (b) A one-man bobsled takes a bend at high speed, also at the 2009 World Cup. Photos by Todd Bissonette. Thanks to the U.S. Bobsled and Skeleton Federation for these images.

Table 3.1. Weight constraints in sled competitions

Sled	Min. sled weight (kg)	Max. sled + crew weight (kg)
Two-crew bobsled	170	
2-woman		340
2-man		390
Four-man bobsled[a]	210	630
Luge	25–27	
Single		102
Double		215
Skeleton[b]		
Women's	35	92
Men's	43	115

[a]There is no four-woman bobsled event.
[b]There is only one athlete for this type of sled.

For all three types of sleds, competitions are decided by the minimum times taken to complete the track. For bobsled, all events require four runs; the team with the lowest total time wins. For skeleton, two runs are made. Luge events are decided over four runs for singles and two runs for doubles. The time it takes to complete a run depends upon the track characteristics. At the Turin games in 2006, the Cesana track was 1,435 m long and contained 19 curves; the winning German four-man bobsled team (Lange, Kuske, Hoppe, and Putze) averaged 55.105 seconds for their four runs, whereas the second-place Russian team averaged 55.138 seconds. Thus, the average speed of the winning team was 93.7 kph. The close result can be expressed differently by considering how far behind the Russian bobsled would have been when the winning German team crossed the finishing line: 0.86 m—less than 3 feet.

We know that bobsleds can attain speeds of up to 140 kph on a straight section of track, so evidently the bends are slowing the sleds. At Turin, the two-man winning team (Lange and Kuske of Germany) averaged 55.845 seconds over the same track. This was slower than the four-man race—the difference amounts to almost 20 m—and so weight is clearly a factor in determining time. The winning team in the women's double bobsled (Kiriasis and Schneiderheinze, also of Germany) averaged 57.495 seconds, significantly slower than the men's double time. Recall that the women's sled-plus-crew weight is 50 kg less than the men's (see table 3.1), and you can see that weight influences run time.

Figure 3.3. Skeleton crew. (a) A skeleton athlete throws himself onto his sled at the start of a run. (b) Skeleton athletes—male or female—are different from normal human beings. Who else would choose to careen down an icy slope at high speed with their chin just two inches off the ground? Both photos by Todd Bissonette. Thanks to the U.S. Bobsled and Skeleton Federation.

Figure 3.4. Lugers are propelled feet first. The ends of the flexible runners are between this athlete's calves; by applying pressure with her legs, she can to some extent steer her sled. How she sees where she is going is another question. Photo courtesy of U.S.A. Luge.

The skeleton track at the 2006 games was the same length and with the same number of curves as the bobsled track. The winning time (that of Duff Gibson of Canada) averaged 57.940 seconds per run.[3] This was slower than the bobsled average. Again, the lighter women were a few seconds back; Maya Pedersen of Switzerland won with an average time of 59.915 seconds. The men's singles luge was won by Armin Zoeggeler of Italy with an average time of 51.522 seconds and a speed of 100.3 kph. Luge competitions are the only events that are timed to the millisecond, instead of to hundredths of seconds. Luge doubles are slower because they (and the women's singles) are contested over a somewhat different track (see fig. 3.1); at 1,198 m it is 176 m shorter, so there is less opportunity to build up speed.

Apologies for throwing these run-time statistics at you, scattergun

3. In the 2010 Games, only 2½ seconds separated first and tenth place in the skeleton competition, after four runs.

GALILEO CONFOUNDED? NOT REALLY

Readers who recall their high school physics may wonder how weight can influence speed. After all, didn't Galileo show, centuries ago in his famous Leaning Tower of Pisa demonstration, that different weights fell at the same speed? More accurately, Galileo showed that two weights dropped from the same height accelerated at the same rate; because both started at zero speed, they would each have the same speed at any given instant of their fall.

In fact, the Leaning Tower of Pisa experiment is probably just folklore. Galileo did indeed play around with different weights—observing how they rolled down inclined planes—but did not actually drop the weights from the Pisa tower. He was trying to demonstrate that the accepted view at the time, due to the Greek philosopher Aristotle, was wrong. Aristotle claimed that the speed of a falling weight was proportional to its mass, so that a 10-pound weight should fall ten times faster than a 1-pound weight. This is not the case, as Galileo readily showed. (Aristotle was a philosopher, not an experimental scientist, so he did not bother to check his assertion.) Galileo's experiments involved dense weights for which he could neglect the effects of friction. That is, the air resistance and the force of rolling friction of a weight passing down an incline are small compared with the force of gravity that acts upon the weight. So, Galileo found that the acceleration and speed of an object that is falling, or on an incline, does not depend upon the mass/weight of the object.

True, in the absence of friction. In a vacuum, for example, a feather and a rock will fall at the same rate. *But* if frictional effects are significant, as for our sleds, then mass does matter, as we will see. The effect is small but—in a sport where hundredths of a second mean the difference between hero and zero—very significant.

---

fashion, but sledding events are all about time. Everything that is done in these events—making the sled more aerodynamic, with smoother runners, speeding up the start, practicing boarding, steering into and out of the curves—is directed toward reducing time, shaving off a hundredth of a second here and there. You should by now be getting a feeling for how sled weight and speed influence run time. To see *why* things work this way, we need to turn to the physics of sledding events. A technical analysis will enable us to understand the results just presented and will provide insight

LUGERS AT WHISTLER

Just before the opening ceremonies of the 2010 Winter Olympics in Vancouver, a fatal accident occurred on the Whistler Sliding Centre track. A 21-year-old Georgian luger, Nodar Kumaritashvili, tragically lost his life at the end of a practice run, when he flew off the track and struck a metal post. Olympic officials decided that, in the interests of the athletes' safety, the start of the men's luge events should be moved to the women's start position. The perceived problem was that the Whistler track was very fast because of its 152-m drop from top to bottom, and too much speed was developed by lugers. (Interestingly, the bobsled and skeleton start positions were not moved.) The last-minute change of start position threw out much of the benefit that lugers had gained by practicing from the original start position. For example, a luger would enter a given bend at lower speed and so would fly up to a lower height around the bend, thus changing the dynamics of negotiating the bend. Canadian lugers, whose home track this was and who had practiced on it the most, did not reach the podium during the 2010 Olympics.

into the subtleties of sled design and sledding technique that mean the difference between first and second place—or the difference between finishing and flying off the track.

SLED PHYSICS: STRAIGHT DOWNHILL

In technical note 1 we see that there are three external forces that act upon a sled:

1. gravity, acting vertically down;
2. contact friction, acting to oppose the sled velocity; and
3. aerodynamic drag, also acting to oppose velocity.

We might add a fourth force: the muscle power of the sled athletes, when initially propelling their sled. And a fifth force: the "normal," or reaction, force of the track pushing back on the sled, because the magnitude of this force influences contact friction, as we saw in chapter 1.

Let's begin by considering an idealized sled track: it is dead straight and with a constant slope. Such a track is ideal for a scientist who wants to analyze sled physics, but is not ideal for the athletes concerned or for watching sports fans. There is little in the way of skill involved for racing on this track (for example, none of the steering skill that is required to negotiate a bend) and little variation from one run to the next. This is the whole point, however, for the scientist, and it will serve a purpose. This astoundingly boring straight track will permit me to show you in a clear manner just how far the physics of sledding can be readily understood, and to demonstrate how the different parameters that define a track (and a sled that slides down it) influence the all-important run time.

To begin, I need to fix the details of those forces. First, gravity is well understood at the level of classical mechanics that we will need here: the acceleration due to gravity at the earth's surface is about 9.81 ms^{-2} acting straight down.[4] Second, contact friction is well-described as being proportional to the "normal force" as we see in technical note 1; the constant of proportionality is called the friction coefficient. We are concerned here with the kinetic friction coefficient between sled runner and ice. This is a matter of considerable interest to bobsled, luge, and skeleton athletes who run down real tracks and not the idealized track I will be considering here. A small difference in friction coefficient makes a significant difference in run time. As you might imagine, sled designers have gone to great effort to design runners that minimize the kinetic friction coefficient.

Twenty years ago, this friction coefficient (denoted μ) was determined to be anywhere in the range of $\mu = 0.01$ to $\mu = 0.05$. The high value is typical of any smooth steel structure sliding over ice; the lower value is achievable by careful design of the runner shape and surface. It used to be the case that a bobsled designer could choose the material of construction for his sled runners and coat these runners with whatever he felt would make them more slippery, but now the rules of the FIBT are very strict. Runners must be round in cross section and made of uncoated, homogeneous solid steel. Recent changes in the rules show that even this restriction is not sufficient: the steel must be that which has been made under the auspices of the FIBT. Even the temperature of the steel runners is restricted; it is forbidden to heat up the runners before a race (hot runners slide better). The rules greatly restrict sled designers' options: runner

---

4. The acceleration at the earth's surface due to gravity varies by 1%–2%, depending upon altitude and latitude.

dimensions are also tightly constrained. Researchers don't yet entirely understand why one set of runners has a lower friction coefficient than another set; consequently, good runners are valuable and are jealously guarded. The friction coefficients of good runners are generally quoted at the low end of the range given above: I will adopt the value $\mu = 0.014$.[5]

Third, aerodynamic drag increases with speed, as we saw in chapter 1. This is basic physics, and the sled designer just has to live with it. But it's an uneasy cohabitation: designers do their utmost to minimize the effects of drag, by minimizing the drag coefficient, c_d, which depends in a complicated way upon the shape and construction of the sled. Bobsled designs, in particular, are expensive (adopting technology borrowed from NASCAR and Formula 1 racing). Bobsleds cost upwards of $30,000—that would have been a couple of houses in Detroit during the 2008–9 housing recession[6]—and much of the design effort goes into drag reduction because, let me say it again, a small reduction in drag coefficient produces a significant reduction in run time. Again, FIBT rules severely constrain the allowed shapes and materials that can be used for bobsled, luge, or skeleton design; thus, bobsleds must be convex in shape with no holes. Designers must work within these rules to come up with a sled that beats the competition. Wind tunnel tests on modern bobsleds show that, for example, a two-man bobsled plus crew will exhibit a drag coefficient of about $c_d = 0.4$; without the crew the figure drops to $c_d = 0.3$. This difference is enormous and shows that crew position, clothing, and helmet are all-important. Analysis of bobsled runs reveals that a drag reduction of 3% translates into a run-time reduction of 0.1 seconds—a huge saving.[7]

In technical note 6, I analyze these three forces as they apply to a sled on our idealized straight track. Here, first the pictures, then the explanations. Figure 3.5 graphs the results of the technical note analysis for track parameters that are close to those of the Cesana track, for which I have already quoted winning run times.[8]

---

5. For contact friction coefficient measurements see Avallone, Baumeister, and Sadegh (2006); Lewis (2006); and van Valkenburg (1988).

6. According to a Reuters online report (Reuters 2009).

7. For drag coefficient figures, see Lewis (2006) and Motallebi, Avital, and Dabnichki (2002).

8. For example, I chose a track length $L = 1,435$ m for my idealized track because this is the length of the Cesana track. Also, the Cesana track drops a total distance of 114 m and thus has an average slope angle of 4.54°; this is the angle I chose for the idealized track slope.

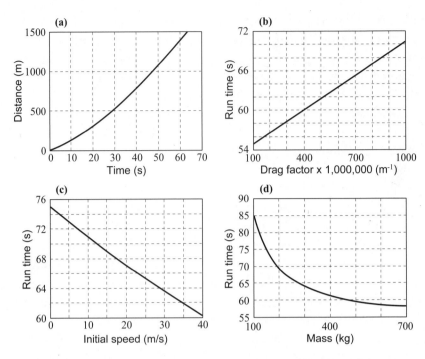

Figure 3.5. Straight track dynamics. (a) Distance traveled vs. time. (b) Run time vs. drag factor, b. (c) Run time vs. initial speed, v_0. (d) Run time vs. mass.

Figure 3.5a shows how the distance traveled down the idealized straight track increases with time: the sled is accelerating. Toward the end of the run, the slope of this curve is becoming nearly constant. This means that the speed of our sled is reaching its maximum value, called the *terminal speed*. Terminal speed is a consequence of drag. Initially, the sled's movement is held back by the friction force, but as speed increases, drag becomes more important. So in the first part of the run, the contact friction force is greater than aerodynamic drag; in the later stages of the run it is drag that dominates. There is a point at which the two dissipative forces—contact friction and drag—balance the downhill force due to gravity, and from this point onward the sled does not accelerate; it has reached terminal speed. This terminal speed is calculated in technical note 6; it is larger for steep slopes and a small drag factor, unsurprisingly.

Figure 3.5b shows us how, on a straight track, the run time changes with the drag factor. In this book drag factor is denoted b; we saw in technical note 3 that it is related to the drag coefficient c_d, cross-sectional area A, air density ρ, and sled-plus-occupant mass m via the equation $b = c_d \rho A / 2m$. Anything the sled designer can do to reduce drag factor will reduce run time. In a sport where hundredths of a second count (thousandths in the case of skeleton) any small reduction matters. This is why such effort has been put into reducing the drag coefficient, c_d, and why ballast is added to bring the mass up to the maximum permitted level: both of these reduce the drag factor, b. This applies for real tracks as well as for the idealized track of technical note 6.[9]

Figure 3.5c shows how run time depends upon initial speed. We will discuss initial speed—the speed that the athletes provide during the push-off stage of the run—in the next section. It is reckoned that initial speed goes a long way to determining the winner, which is why all sled athletes (particularly bobsledders) need to perfect their technique; they want to get up to maximum speed and then get aboard the sled as quickly as possible to reduce drag and increase mass. The graph in figure 3.5c applies in detail only for my hypothetical straight track, of course, but the general conclusion applies also to real tracks.

Figure 3.5d shows how increased mass also reduces run time. Again I can apply a reality check. For the numbers presented earlier for men's and women's bobsledding, we find that a 50 kg difference in mass led to a 1.6-second difference in time. The calculation of technical note 6 shows that most of this difference (1.25 s) is due to mass. The rest of the difference is due to individual sled and driver characteristics, and to the varying shapes of tracks (the Cesana track and the Whistler track of fig. 3.1 are anything but straight).

Our analysis of the idealized straight track tells us that the basic physics of sliding on ice, and of aerodynamics, determine much of the characteris-

9. To appreciate the narrow margins of victory in modern bobsled competition, consider the 1998 Winter Olympics in Japan; in one event Great Britain and France tied for bronze medal with exactly the same run time. See Hubbard (2002) for an accessible account of the technological efforts to achieve optimum bobsled design. Note also that the equation obtained in technical note 6 for the terminal speed of sleds is close to measured values (about 140 kph), given the maximum sled mass and measured drag coefficient, if we assume that the cross-sectional area is that of the track rather than that of the sled (typically 140 cm wide, with a height of 50 or 60 cm). This makes sense in terms of fluid dynamics: the airflow is not free but is constrained, almost as in a wind tunnel: we must consider the effect of the track walls when considering air flow.

tics of sledding events. Sled athletes must wear smooth suits and specially designed helmets to reduce drag. They must give their sled the maximum possible initial speed during the push-off stage, get aboard as quickly as possible, and adopt positions that minimize drag. Their design of their sled must be aerodynamic and its runners must be as smooth as possible. Weight (mass) must be as close to the allowed upper limit as possible.[10] Any remaining difference in performance is due to athlete skill in getting up to speed and in steering his or her sled down the track. We now turn to the first of these skills—the push-off.

SLED PHYSICS: A FLYING START

When the green light tells a bobsled crew to start their run, they begin by pushing the sled with all their might. After the push-off, which takes up about 15 m, the crew clambers aboard quickly and adopts their run positions (to reduce aerodynamics drag); they board as smoothly as possible so as not to rock the sled (which would increase contact friction). When they have traveled about 50 m, they cross the start line and a clock begins timing the run. (The time between push-off and the start line—let me call it the "push-off time"—is recorded for interest, to see how fast the push-off was achieved, but the run time begins at the start line.) Clearly, it is advantageous to be moving as fast as possible by the time the clock starts; a widely quoted rule of thumb states that 0.1 second shaved from the push-off time means 0.3 second taken off the run time.

For skeleton events there is a similar push-off phase: the athlete pushes or carries his sled and then launches himself upon it and adopts the most streamlined position. Luge starts are different; we will look at these later on.

By applying the laws of physics to the push-off phase of a bobsled or skeleton event, we can estimate the power that each athlete must exert to achieve a given start speed. Put another way, we can determine the start speed that will result from a given amount of push-off power. Those of you who are following the math can check the analysis in technical note 7. Results are shown in figure 3.6.

Measurements of push-off times from two-man bobsled events show that the push-off exertions of sled crews last for about 5 seconds. Athletes

10. Muscle is more useful than ballast, and so bobsledders tend to be big people.

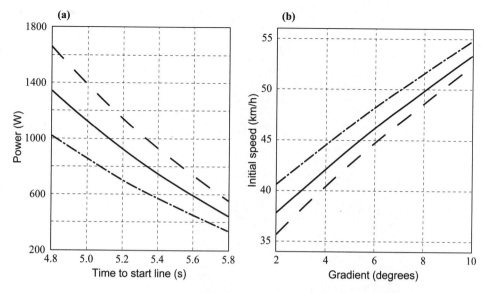

Figure 3.6. Bobsled and skeleton push-off phase. (a) Power per athlete vs. time to start line, for an initial gradient of 5°. (b) Initial speed vs. gradient, assuming that each athlete exerts 1,000 W of power during push-off. Curves indicate four-man bobsled (solid line), 2-man bobsled (dash), and skeleton (dash-dot).

can achieve quite high power output over such a short time span;[11] the power required to reach the start line is plotted in figure 3.6a. Note that a big increase in athlete power exertion is needed to cause a small (but significant!) reduction in push-off time. Push-off time does not contribute directly to run time, but you can see that covering the 50 m distance to the start line in a shorter time means that you begin the event at a higher speed. In figure 3.6b you can see that this start-line speed increases with track gradient, naturally, and that it varies between events (four-man bob, two-man bob, and skeleton). We saw in figure 3.5c how run time was reduced by increasing start speed; by glancing at both figures 3.5c and 3.6b you can appreciate how an athlete's greater effort at the start pays off in

11. For example, if you run up five or six steps in one second, you are doing work at a rate of about one horsepower (1 hp = 746 W). Note that power exerted is different from power applied because you are not a perfectly efficient machine. Only about 20% or 25% of the energy that you expend upon a mechanical task actually contributes to work done. The rest is wasted as heat, or sound, or is used internally to increase heart rate and so on. See Williams (2001) or Komi (2002) for details of human power exertion in sport.

Figure 3.7. Bobsled start. (a) Two-woman bobsled push-off and (b) getting aboard. Both aspects of the bobsled start must be fast if the run is to lead to a podium. Photos by Todd Bissonette. Thanks to the U.S. Bobsled and Skeleton Federation.

terms of reduced run time. Bobsled athletes must push off very hard and then get aboard their sleds as quickly as possible (fig. 3.7).

The luge push-off is shorter in both time and distance, and it results in a correspondingly smaller initial speed. A luger starts by sitting on the luge and pulling hard on start handles that are attached to the side of the track. This initial effort involves only upper body strength. Then he or she vigorously paddles the ice three or four times, with spiked gloves, to add to the push-off speed. Thereafter gravity and friction take over.[12] In technical note 7, I provide a rough back-of-the-envelope calculation, demonstrating that the initial speed generated by the pull increases as the square root of power. Thus, a slight increase or reduction of applied power results in a slight increase or reduction of initial speed. In technical note 7 the results of technical note 6 are applied and show how such a small change in initial speed influences run time. Of course, the calculation in technical note 6 was for an idealized straight track and so we must not take the numbers too literally. Bends and varying ice conditions such as are found on real tracks will also influence run time. However, we are safe in concluding that the straight-track result is pointing us in the right direction, so to speak: a small change in luge initial pull power makes a measurable difference in luge run time.

SLED PHYSICS: DRIVING YOU ROUND THE BEND

Now I get real by allowing our sled track to have bends. In bobsled, luge, and skeleton competition events, it is the bends that make the track; they define which tracks are easy and which are hard, they cause the spills and thrills, and they result in steering errors that go a long way to deciding medal places.[13]

We are all familiar with the centrifugal force that arises when we take a corner at speed. Consider figure 3.8. First, imagine that the sled is a flat-bottomed toboggan (no runners). Apart from the dissipative forces of

12. A description of luge and of the work performed by lugers can be found in, for example, Garrett and Kirkendall (1999).

13. Thus, the Cesana track in Turin is considered difficult because of the technically challenging "Toro" series of bends near the beginning. The Whistler track, as we have seen, is very fast—perhaps too fast. There are only 16 bobsled tracks around the world that are recognized by FIBT (they are very expensive to build), and each is characterized by the combination of bends more than by the length or gradient; sled crews must practice on the track prior to an event, to learn its secrets.

drag and friction (which I will set aside for now) there are two external forces acting upon the toboggan: gravity, as always, and now the centrifugal force, which acts to throw the toboggan off the bend. In figure 3.8a we have a bend of radius R; the centrifugal force acts along a radial line, out from the center. If the track is level (not banked), as shown, the centrifugal force will cause the toboggan to slide off the track. Suppose now that we are in a bobsled with runners. The runners will resist slipping sideways, just as ice skates do, and so will resist the centrifugal force, but at a cost of increased contact friction. (If the centrifugal force is strong enough, the bobsled may flip over.)

Let us say that you are a bobsled driver and you try to steer a course along the middle of the track shown in figure 3.8a. You find, however, that the sled wants to fly outward, even though the runners prevent this from happening. This is because the torque that is applied by the centrifugal force—trying to tip over the sled—leads to an increased weight on the outer runner and a reduced weight on the inner runner. Recall that increased weight means increased friction force; thus, the outer runner experiences greater friction on the bend than does the inner runner. This frictional asymmetry causes the sled to turn toward the outside of the bend; it is as if you were dragging one foot to steer the sled. So, even with runners preventing a sled from sliding outwards, the centrifugal force will cause the sled to steer outwards. This trend can be resisted by you, the driver, but it must be done carefully to avoid flipping the sled.

The centrifugal forces that act upon sleds are so large that, of course, real sled tracks are very steeply banked. The section of track in figure 3.8b is always taken at high speed, and so it looks like it has been built on its side. A halfpipe track looks like half a cylinder: the bank angle is zero (horizontal) in the center of the track and increases to 90° (vertical) at the outside edge. Because of this wide range of banking angles, there is a "natural" place on such a track for the sled to find itself when flying around a bend.

In technical note 8, I show how the position of a sled on the banked curve depends upon sled speed (from everyday experience you know that faster sleds will need to be further up the bank). Technical note 8 also shows you how the frictional force increases on banked curves (on *both* runners); friction increases so much that sleds are significantly slowed when taking bends at high speed. The g force experienced by sled crews increases from 1 g on the straight (i.e., the usual downward force of gravity)

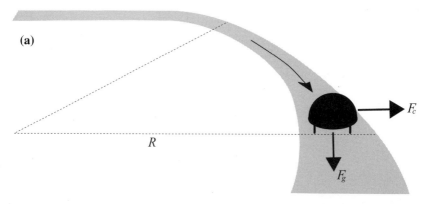

Figure 3.8. Centrifugal force effects on a sled. (a) When speeding around a bend of radius R, the sled experiences a centrifugal force F_c in the direction shown, as well as a gravitational force F_g, directed straight down. (b) Because of centrifugal force, high-speed bends are steeply cambered. This one is part of the track at Lake Placid, New York. Photo by Todd Bissonette. Thanks again to the U.S. Bobsled and Skeleton Federation.

to 4 g or 5 g on a tight, high-speed bend.[14] Frictional force is increased by the g factor, as you can see in technical note 4. It is as if the kinetic friction coefficient has suddenly increased from $\mu = 0.014$ to $\mu = 0.06$ or 0.07.

14. FIBT rules restrict the maximum g force that is experienced by sled crews to 5 g, and this only for no more than 2 seconds.

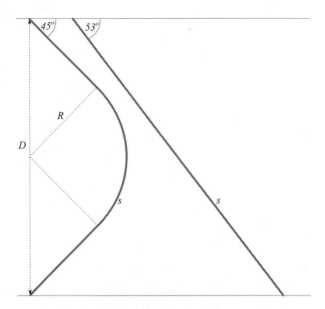

Figure 3.9. Two tracks (bold lines) of the same length, *s*, down a constant-gradient slope. The height difference between the two contours (horizontal lines) separated by 100 m is 8.7 m. Because of increased friction around the bend (radius *R*), the curved track is slower than the straight track.

Look at figure 3.9: here we have two tracks of the same length with the same downward slope. The track with the bend will be slower than the straight track (by about 0.03 s), even if all other conditions are equal, because of the effect of increased friction due to centrifugal g force, as explained in technical note 8. Now multiply up the section of track shown in figure 3.9 so that we have a track length of 1,500 m containing, say, 12 bends: such a track will be about 0.4 second slower than a straight track of the same length. I need not waste words emphasizing to you the enormity of this time difference in sled competitions.

There is more to sled steering than simply avoiding being overturned by the centrifugal force around bends, of course. Consider the halfpipe track bend illustrated in figure 3.10. The black lines show where a stable sled should be positioned. On the straight sections it should be at the bottom of the pipe. On the curved section it should be higher up the outside part of the bend. In reality, a sled cannot simply jump from the bottom of the bend to some other location; it must gradually climb the bend, and then gradually descend once the bend straightens out (as shown by the dotted line in fig. 3.10). During the transition between bend and straight section, the sled will not be at the "natural" bank angle. For example, when coming off a bend of radius R into a straight section of track, the sled will be high up the bend, but as soon as it crosses from bend to straight section the centrifugal force is suddenly switched off and the sled is high and dry (well, high anyway). It will fall down to the bottom of the pipe unless the driver carefully steers it gradually down the slope. Just how gradual the steering needs to be depends upon the track (there may be another bend

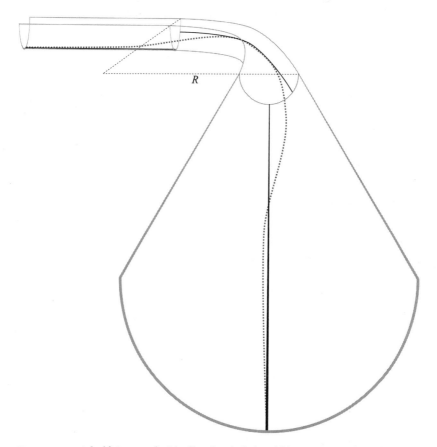

Figure 3.10. A halfpipe track. Ideally, the sled should be positioned on the black lines: at the bottom of the halfpipe track during straight sections and at some high angle up the bank during the bend. In reality, this is not possible; the sled must make the transition gradually (dotted line). This means that inevitably the sled cannot always be in the ideal position. The result is that it experiences sideways forces and increased friction.

just ahead). Decisions just like this—where and when to steer into and out of a bend—make the difference between a gold medal-winning sled driver and one who finishes in second place or in the hospital.

Imagine the difficulties faced by a bobsled driver as he descends an Olympic track such as that at Whistler, or Lake Placid, or Cesana. He is inside a box that rattles noisily as it whips around the course at upwards of 140 kph. He must keep his head down to reduce aerodynamic drag while trying to anticipate the next bend. He shoots around the bend and feels extra g forces throwing him to left or right. The friction impacting his sled is suddenly increased and equally suddenly drops. He must steer a course through a bend that will keep him inside his sled and will shave hundredths of a second off his run time, while being subjected to a g force that makes him feel as if he weighs 900 pounds. Now imagine descending the same course in a skeleton: There are no comforting side walls; your chin is 2 inches from the track; your arms are by your side (which must be psychologically difficult—I would want to hold my arms in front of my face); and you can steer only by carefully shifting your weight. A luger descends the same track on her back, looking between her feet to see the bends ahead; she steers with her calves while being rattled around like a pea on a drum. Not for me. I will be staying on my comfortable sofa.

4   PUCKS AND ROCKS

Intrepid winter sports athletes hurl themselves at high speed over ice: they aim to slip, slide, slither, and glide their way from A to B in the fastest time, or at least faster than their opponents. Two of our winter sports, however, involve athletes hurling objects other than themselves. Sure, a hockey player is happy to turn himself into a projectile, directed at an opponent or at a location on the ice where he anticipates that the puck will be, but he will also send the puck across the ice or through the air. A curler sends a 40-pound granite rock over 100 feet along a sheet of ice, at a much more sedate pace. In this chapter we will investigate the unusual—indeed, decidedly odd—motion of ice hockey pucks and curling rocks.

PUCKS

A modern hockey puck is a short cylinder made of vulcanized rubber. It is a quite hard form of rubber but is not brittle; it deforms, instead of shattering, when struck. The cylinder has a diameter of three inches and is one inch thick. Pucks weigh about 6 ounces (170 g). A puck is a pretty mundane object to look at, but it attracts great excitement in a hockey fan by virtue of its position, when it flies at great speed towards his face or when it crosses a goal line. To a physicist, however, the humble puck is a very interesting object even when moving at low speeds. I will look at the physics of pucks at both extremes of speed, and begin with one that is sliding over the ice very slowly.

The Puck Stops Here

A puck moving along the ice will, in general, be spinning about its center, as well as sliding. Spin is imparted by a hockey stick when the puck is struck, as we will see, and much of this angular speed is retained throughout the subsequent trajectory of the puck. Consider now a puck that rotates as it slides over the ice. The particular trajectory that I will consider here is one we don't see all that often in hockey: the puck is allowed to come to a stop naturally, without interference from a player.

The physics of puck motion is governed by sliding friction, with a friction coefficient of about 0.015. Theory predicts the following odd behavior: the puck stops moving over the ice at the same time as it stops spinning. It doesn't matter how much speed or spin the puck starts out with—the two types of motion stop together. This prediction is borne out by experimental observation. It applies to any circular object sliding over the ice. The details differ from one object to the next (depending mainly upon the sliding object's mass distribution), but the result is the same: spinning and sliding cease simultaneously.[1]

This strange behavior was investigated a quarter century ago and is still generating interest in physics journals.[2] It seems that the reason why spinning and sliding stop at the same time is because the two types of motion become tied together by the force of friction. The exact way in which they are connected depends upon the slider's mass distribution, but it can be calculated. For a puck, for example, we know that as the sliding speed slows, spin rate slows down as the square root of speed. This means that during the time it takes the puck's speed to drop to a quarter of its initial value, the spin will decrease to half its initial value. When the speed has been reduced to one-ninth, the spin will have been reduced to one-third. At the very end of the puck motion, just a fraction of a second before it stops, the spin and speed exhibit different behavior; they are linked, but in

1. The same odd behavior may apply to sliding objects that are not circular, but I am not aware of any theoretical or experimental investigations into the sliding behavior of such shapes. By "mass distribution" I mean the variation of slider mass with radius. A puck has a uniform mass distribution—it is the same at all radii—whereas the mass distribution of a hemisphere sliding with the flat surface against the ice has a mass distribution that decreases with increasing radius. Such a hemisphere, even if it has the same total mass, contact radius, and friction coefficient as a puck, will slide differently in detail, but like the puck it will stop moving and spinning at the same time.

2. See Voyenli and Eriksen (1985) and Denny (2006).

a different way. At this late stage spin slows down in a manner that is proportional to speed.

Curling rocks slide over the ice and are circular, so we can expect that they, too, exhibit this association of speed and spin, and in fact this happens. The details are different, however, as we will see.

So what are the consequences, for the game of hockey, of this unexpected physics displayed by the humble puck? None at all, so far as I can see. I just mention it because I am a physicist as well as a sports fan.

Slap-Shot Science

The other extreme of puck speed is of great importance to the game of hockey, however. A hockey player can accelerate a puck via wrist shots, snap shots, sweep shots, and slap shots. It is the last of these that will concern us here because slap shots generate the fastest pucks. In a slap shot, the hockey stick strikes the ice just before hitting the puck. The stick bends; it is elastic. This bending is quite marked and can easily be seen in photos of slap shots.[3] Bending happens even before the stick strikes the ice, as a result of the force applied by the hockey player when swinging the stick, but the ice causes it to bend more. Being elastic, the stick rebounds—recovers its shape—but the rebound causes the blade to move faster through the air, toward the puck, than would be the case if the stick had not struck the ice. This behavior is what gives the slap shot its name and the puck its maximum speed.[4]

Interestingly, a study shows that female hockey players could produce faster slap shots if the hockey sticks they used were a little less stiff—a little easier to bend. This is because female players are not generally strong enough to bend a normal stick (hockey stick design is based upon male players' height and strength), and therefore do not benefit as much from the elastic effect just described.[5]

How much speed is imparted to a puck that is on the receiving end of a slap shot? Slap shots by professional hockey players have produced puck

3. See, for example, the front cover of Haché (2002).

4. For technical details about slap shot physics, see, e.g., Lomond, Turcotte, and Pearsall (2007).

5. For more on the improvement of women's slap shots with more flexible hockey sticks, see Gilenstam, Henriksson-Larsén, and Thorsen (2009).

Figure 4.1. The fastest team sport. Note the short skates and curved blades (on the hockey sticks as well as the skates). Thanks to Brooke Novak for this image.

speeds in excess of 160 kph (100 mph).[6] This means that the puck has been jolted with about 170 J (about 125 ft-lb) of energy, which is enough to kick a hockey player 8 inches (20 cm) off the ice. At these speeds, the puck becomes a dangerous missile, as the padding and face masks of goalies testify. Even spectators are at risk from a wayward shot.[7]

In the 1960s players used to bend the blades of their hockey sticks because they thought that it made the resulting slap shot more difficult for a goalie to stop. This practice was eventually restricted. In the NHL nowadays a player is allowed to bend a 32-cm blade no more than 2 cm (see fig. 4.1); any more results in a penalty for the player found to be carrying such a stick on the ice. The flight of a puck became erratic when it was fired

6. Bobby Hull, in the 1960s, was an exception. His slap shot was measured at a stunning (quite literally, for some goalies on the receiving end) 120 mph. See Haché (2002, 87–90).

7. Spectators who have been hit in the face by a fast-flying hockey puck have wound up with broken jaws, and with tooth and nerve damage. In Germany this has led to the placement of safety nets around the ice (thus obscuring the view of fans and TV cameras). One study shows that the risk to spectators can be reduced by 80% if security glass 3.8 m high surrounds the ice; it would have to be 10.8 m high to eliminate the risk entirely! See Böhm, Schwiewagner, and Senner (2007) for details. For more on the impact of a fast puck upon goalies' face masks, see Castaldi and Hoerner (1989).

from a stick with a curved blade. Also it is likely (though I can find no figures to prove it) that the puck flew off the stick a little faster. Both of these effects would make the puck harder for a goalie to stop. Can we understand why bending a hockey stick blade would have such effects on puck flight? I think so, although, again, I have no evidence to back up my reasoning. Regard the next couple of paragraphs as speculation, awaiting experimental confirmation or refutation.

First, the extra speed. A curved blade probably means that the puck remains in contact with the stick blade for longer, during the slap shot. That is, the short interval from the first contact to the moment the puck separates from the stick is probably longer if the stick blade is curved. You can see why in figure 4.2. A straight blade strikes the puck and causes it to deform. The deformation is eliminated as the elastic puck reverts to its original shape and bounces off the stick. For a curved blade, the natural sweeping action of the slap shot will cause the blade to remain in contact a little longer. As the puck reshapes from its squashed state immediately after being struck, it rolls up the surface of the blade (and so is given a spin as well as a little extra push). The extra contact time means higher speed, even if the curved blade and the straight blade strike the puck with the same force. This is because puck speed increases with imparted energy, and energy is the force of the slap shot multiplied by the distance over which this force applies—which, we are saying, is likely greater for a curved blade.

What about the erratic flight of a puck off a curved blade? The puck will

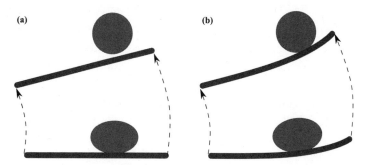

Figure 4.2. Slap shot. (a) A puck is struck by a straight blade. It is deformed by the force of the shot before recovering its shape and bouncing off the blade. (b) A curved blade may be in contact with the puck for longer and so impart more energy to it.

be given more of a spin by a curved blade than by a straight one, as you can see from figure 4.2. This means that the puck may curve more in flight due to the Magnus effect—the same phenomenon that causes a spinning baseball to curve sideways. Also, because more energy is deposited in the puck by a curved blade, it will deform more; this deformation may influence the aerodynamics of the puck flight, at least during the short initial stages of its trajectory as it leaves the blade. It would be difficult to predict in detail how a puck that is accelerating, spinning, and changing shape (as it recovers from deformation caused by the slap shot) behaves as it travels through the air, but it is not hard to imagine that the flight would be erratic.

ROCKS

The rocks used in curling, called "stones" in Europe, where the sport originated, are pill-shaped 18-kg granite blocks, with a handle on top (fig. 4.3).[8] If you are not familiar with the sport of curling, then there are a couple of facts about it—about the curling rocks that are thrown during the game and about the ice over which they travel—that will surprise and amuse you. So it will be worthwhile for me to explain to you the basics of the game.

Curlers, four to a team, each launch two rocks down the ice at a target. The target consists of concentric circular rings, the outer ring having a diameter of 12 feet. The *button* (the bull's eye) is 126 feet (38 m) from the *hack*, from which the rock is thrown. The curler pushes off from the hack, sliding down the ice toward the target area (the *house*), with the rock held in front (see fig. 4.4); just before it is released, the rock is given a twist, either clockwise or counterclockwise. Players take turns sliding a rock along the ice; once all eight rocks have been played, the team with the rock closest to the button gains a point (or two points if they have two rocks closer to the button than any opposition rock, three points if . . . etc.—you get the picture). There are a lot of tactics involved in the game: a rock may

8. Some people say that the game of curling started in Scotland; others insist that it was brought there from the Low Countries. Readers may be familiar with the painting *Hunters in the Snow* by the sixteenth-century Dutch artist Pieter Breugel, which includes, as part of the background detail, a game of curling. There are also sixteenth-century Scottish references to the game. Take your pick. What is more certain is that the modern rules of the game originate in Scotland.

Figure 4.3. Curling rocks. The handle is used to push the rock along a sheet of ice and to impart a twist to it, which causes the rock trajectory to curl. Thanks to Jerome Larson, USA Curling, for this image.

be sent down the ice to knock opposition rocks out of the house; another may be intentionally left short of the house to guard against such *take-out* shots.

The interesting physics of curling centers on the feature that gives the game its name. A rock that is sent down the ice rotates slowly as it slides. This rotation imparts a curve to the trajectory. That is, a rock that is given a clockwise spin will develop a curve to the right, as seen by the curler who launched it, whereas a rock with a counterclockwise spin will curl to the left. Just why this happens is still something of a mystery, which I will address soon enough. The mystery deepens when we consider the ice upon which the game is played. Curling ice has to be very flat—flatter than ice used in a hockey game or for skating competitions. Very flat ice is required so that the rocks do not slide down any slight slope that may otherwise exist; the only curl in the rock trajectory should be due to the spin, and not due to a slope on the ice. The strange fact about curling ice is that it is *pebbled*. Part of the preparation of a sheet of ice for the game of curling involves sprinkling water droplets onto the flat ice; these droplets freeze to leave a dense covering of ice pebbles—raised bumps over which the curling rock travels. Thus, curling ice is globally flat but locally very bumpy. Without these bumps—on ice that is glassy smooth—the rocks curl very little or not at all.

The plot thickens when we turn a curling rock on its side to examine the bottom surface. It is not flat. There is a raised ring, an annulus with a diameter that is about half the rock diameter, which is the only part of the rock that is in contact with the ice. This ring, known as the *running band*, resembles the ring on the bottom of a dinner plate or a teacup. If the rock had a flat bottom instead of this running band, it would curl differently. Pebbling and the running band are shown in figure 4.5.

Before addressing the curling mystery, there is one more facet of the game that needs to be aired. A curling rock slides over 100 feet (30 m) from the point of release to the position at which it comes to rest (unless it is a take-out shot, in which case it strikes another rock and careens out of the ice sheet). Between these two positions, physics determines what happens to the rock. However, there are two *sweepers* who can influence the

Figure 4.4. Launching a rock. The curler slides down the ice, aiming the rock in the direction specified by the team's skip. The rock is given a twist just before it is released. Thanks again to Jerome Larson, USA Curling, for this image.

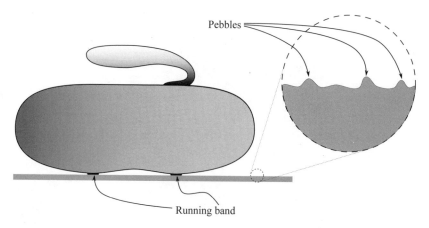

Figure 4.5. Rock and ice. A raised ring called the running band is the only part of the rock that touches ice. Curling ice is pebbled—that is, it is provided with numerous raised bumps over which curling rocks slide.

motion of the rock. Sweepers do what the name suggests: they use brushes to sweep the ice that is immediately in front of the rock. They do this, not out of some innate sense of tidy housekeeping, but because vigorous brushing of the ice changes its characteristics and therefore changes the rock trajectory. In particular, sweeping reduces the sliding friction coefficient so that the rock will travel further. Thus, sweeping can increase the distance traveled by a rock that looks to be *light*—one that will otherwise finish short of its desired position. Sweeping also reduces the amount of curl, and so a rock that looks like it is going to curl too much may also be swept. A rock that is *heavy* (given an initial speed that is too great), or one that needs to curl a lot in order to end up in the right position, is not swept. It is reckoned that effective sweepers can change the final position of a rock by several meters (see fig. 4.6).

There is another reason for sweeping (lightly, in this case) the ice in front of a moving rock: to remove grit or other debris that inevitably accumulates on the ice. Some of this grit can profoundly alter a rock trajectory; it can cause the rock to veer unexpectedly or *reverse handle* (switch from, say, a clockwise to a counterclockwise twist). This is obviously undesirable, and so ice may be lightly swept for this reason alone.

The ice must be swept just before the rock passes over it. Experienced and expert curlers suggest that a second rock sent along the same trajectory

Figure 4.6. Sweeping a rock. This is the only part of the game where physical strength matters, and so men tend to be more effective sweepers than do women. To be effective, players must lean heavily on the brush and sweep it across the rock path vigorously. I am grateful to Jerome Larson, USA Curling, for this picture.

as the first, *immediately* after the first, will need very little or no sweeping because the ice is still in its "altered state" from sweeping.[9] However, wait a few minutes and this trick won't work: even sent along the same path, the second rock will end up in the same place as the first only if the ice is reswept. So how does sweeping alter the state of the ice, albeit temporarily? It used to be thought that sweeping melted the ice so that the rock could hydroplane over a thin layer of water, but this is no longer considered to be the case. (Calculations show that the layer of meltwater would be much too thin—less than the pebble height.) Nowadays, theory and observation suggests that sweeping increases the temperature of the ice without melting it; this softens the ice and reduces the friction coefficient. It also removes ice debris that may have accumulated on the surface and may play a role in determining rock motion. Clearly, waiting a few minutes will undo these effects (the warmed ice will return to its earlier, colder temperature, and loose ice debris will refreeze to the surface), and the ice will need to be reswept for the next rock.

9. Lino di Iorio, personal communication.

Rock Concert

As with rotating pucks, the spin and speed of curling rocks work in concert because of the forces that are generated between ice and rock. The details are different, however, because the contact surfaces of a puck and a rock are different, and the ice over which they travel is different. Rocks in general lose their spin more slowly than do pucks. (I provide a simple example in technical note 9.) In this case the rock spin slows down as the one-fifth power of rock speed. The speed slows linearly with time so that, for example, after half the trajectory the rock has slowed to half its initial speed; after 90% of the trajectory the speed has decreased to 10%. However, halfway through the trajectory, the rock retains 87% of its initial spin; after 90% of the trajectory, the spin is still at 63%—so most of the spin is lost right at the end.

Any reasonable model of curling rock physics will predict this "concert," this coupled dance between spin and speed, though the details change from one model to the next. (The power may change from ⅕, for example, but it will always be much less than 1.) The link between spin and speed is seen in figure 4.7. (A point on the running band has a local speed over the ice that is the sum—the *vector* sum, technically—of spin speed and center of mass speed.) You may well respond: "OK, you physicists can construct several math models which explain this linked behavior of spin and speed. So what? Only one of them is right. Which one?" Awkward question. It's a good point, however. If several physical descriptions make the right prediction about the way spin and speed interact, then to decide which description is the true one we must look at other predictions that the math models make. In particular, a math model that truly represents the physics underlying curling rock motion must be able to explain the strange curling behavior. So now it is time for us to grapple with that thorny issue.

It Isn't Easy to Bend Rock

In curling it *is* easy to bend rock: every time a curler sends a rock down the ice, it bends. OK, so it is the *trajectory*, not the physical rock, that bends. The amount of bend varies from rock to rock and, in particular, varies from one ice sheet to the next. If a rock travels 30 m down the ice before coming to rest, then it may bend 0.5 m or three times as much. These numbers are typical, but there are exceptions both ways. I have seen rocks

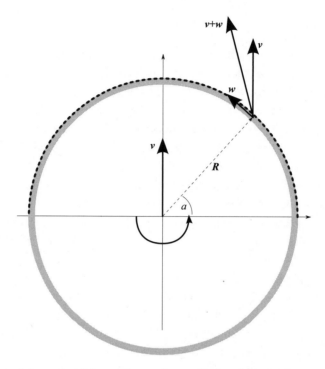

Figure 4.7. Schematic of the curling rock running band. Here, the running band (radius R) of a curling rock is shown, along with the rock velocity vector v and the velocity vector w, due to rock rotation, that applies at angle a on the running band. The local velocity over the ice, at angle a, is given by $v + w$. This vector sum changes with location a around the running band because the direction of w changes. The rock is moving and spinning in the directions indicated by the central arrows. In the impulse model of curling, ice debris accumulates at the leading semicircle, as shown by the dashed line.

on a sheet of ice that do not bend at all, and others that bend 2 m. Here's the strange thing, the oddity that investigating physicists are having difficulty getting their heads around: the curl distance is pretty much independent of the number of rock rotations.

If the rock is not provided with an initial spin, then it will not curl; that's a given. There is no asymmetry—nothing to distinguish left from right—and thus no reason why the rock should move one way or the other. It doesn't; a rock without spin will travel along a straight line. If, however, the rock is given a counterclockwise spin, it will curl to the left by a fixed amount (or to the right for a clockwise spin). That fixed amount is different for different ice conditions, as I have just indicated, but on any given sheet of ice the curl distance (let us say it is 1 m) doesn't depend upon the amount of spin. If the rock rotates once during its trajectory, or 10 times, or anything in between, it will still curl 1 m. I am exaggerating only slightly here; if the rock makes only a quarter of a rotation then, yes, it will curl less than if it makes a half rotation. In practice, in a curling game, players do not provide rocks with such small rotations, however (and probably for that very reason). Curlers ensure that the rock makes at least one complete rotation before it comes to a halt.

Maybe it shouldn't be a surprise to learn that curl distance doesn't depend upon the number of turns, at least for the range of values that curlers typically adopt during a game. (Some curlers will cause their rocks to rotate once or twice as they travel down the ice, whereas others may send off rocks that spin 8 or 10 times before stopping.) It certainly shouldn't be a surprise to anyone who has watched a game, and listened to the team captain—the skip—at one end of the ice shouting to a teammate at the other end with instructions about the shot she is to make. The skip will indicate what line the rock is to follow, usually by positioning a brush in the place she wants her teammate to aim for. She will also indicate what weight is to be used—in other words, how fast the rock should be sent down the ice. These two instructions, for direction and speed, specify the rock's *velocity*. But when it comes to spin, the skip will only say "in-turn" or "out-turn," which specifies the direction of the spin—clockwise or counterclockwise.[10] So, spin direction is indicated, but the magnitude of spin is not. The skip does not shout out "five turns please" or "we need three turns for this shot" because the number of turns (so long as it exceeds one) simply does not matter.

This is very odd. Clearly, providing the rock with spin is essential for it to curl, and yet the degree of curl seems not to depend upon the degree of spin. This bizarre fact seems every bit as weird to investigating physicists as it does to curlers and sports enthusiasts. In fact, it is quite difficult to produce a theory of curling that leads to this prediction. Most theories of

10. When providing a rock with an *in-turn*, the curler's elbow moves toward her body; for an *out-turn* it moves away. Thus, for a right-handed curler an in-turn is clockwise and an out-turn is counterclockwise.

curling that have been produced over the last 30 years have led to the wrong prediction: that curl distance increases with spin—and in particular, that curl distance is directly proportional to the number of turns that a rock makes.[11]

For a math model of curling physics that leads to curling behavior, it is necessary to make use of an asymmetry—a difference between left and right, for example—because otherwise the curling rock will have no reason to curl, as we have seen. The most natural asymmetry to exploit is that of local running band velocity. We saw in figure 4.7 that the velocity at a location on the running band changes with running band angle. It is different on the left and right. The speed (magnitude of velocity) also varies left to right. Add to this suggestive fact the observation that, for many sliding objects, the friction coefficient varies with speed, and you have a natural contender for explaining curling behavior: rocks curl because the friction force varies around the running band. Indeed, it is quite straightforward to produce a theory along these lines (let me call it the "variable friction theory") that leads to curling. Unfortunately, such a theory bumps up against the awkward fact that curl distance is stubbornly independent of the number of turns. This is because the variable friction theory predicts that curl distance increases directly with the number of rock rotations.

Worse, we now know that almost any conceivable variation of the variable friction theory will also produce the same bad result. Whatever is causing rocks to curl, it is not a simple application of the variable friction theory. It is possible to twist theories as well as curling rocks. I have produced an oddball version of the variable friction theory that leads to the correct kind of curling behavior, one that is not sensitive to the number of rock rotations, but this theory only permits curl distances of less than 0.5 m.[12] At best, it is only part of the story. There is some extra physics at play that we do not yet understand.[13]

11. See, e.g., Johnston (1981), Penner (2001), or Shegelski, Niebergall, and Walton (1996).

12. See Denny (2002).

13. In attempting to explain the strange behavior of curling rocks, one group of investigating physicists has even felt the need to ditch a cherished physical principle—that the force of friction acting upon a body always opposes the direction of motion of the body. See Shegelski and Niebergall (1999). Because this principle applies widely—well beyond the physics of curling—it seems unlikely to me that such a drastic measure is necessary. More likely, the reason we physicists are having difficulty explaining curling is because we have, so far, lacked imagination.

ANOTHER CURLING ODDITY

The figure shown here illustrates another example of odd behavior that can sometimes be seen in curling rocks. This time, at least three rocks are involved, and the oddity appears only when two of them are initially stationary and touching. In the figure the left rock (number 1) approaches the other two, which are stationary, before being hit. After the collision between numbers 1 and 2, all three rocks move. Number 1 probably doesn't go very far—it may move north a little. Number 2 squirts out southeast, as indicated by the arrow. We expect number 3 to head northeast, as shown. Most of the time it does, but if rocks 2 and 3 are touching and frozen to each other, then number 3 may head out in a different direction, as indicated.*

Curlers call this phenomenon "the drag effect"—not to be confused with aerodynamic drag. Sometimes ice debris spatters the sides of curling rocks, and if two such rocks are at rest and touching for some time, they can freeze together. The bond is not strong—a collision such as that indicated in the illustration will easily wrench them apart—but it is enough to alter the trajectory of rock number 3. Why does this happen? When two rocks are frozen together, they act briefly like a single body (until the ice bond is broken by the collision). During this brief period, immediately after the collision of rocks 1 and 2, the "combined" rock (let us call it "2-3") heads off in the same direction as number 1, the rock which set it in motion. Then

Curling rock drag. Here, rock number 1 collides with two stationary rocks. We expect rocks 2 and 3 to shoot off in the directions indicated by the black arrows, but if they are frozen together, number 3 instead moves along a direction (open arrow) that is closer to the original direction of rock number 1. This aspect of curling physics, at least, we understand.

2-3 breaks apart, and the two component rocks move separately. The brief union of rocks 2 and 3 is enough to compromise the directions in which they are sent. In particular, rock number 3 moves in a direction that is somewhere in between that of the initial moving rock (number 1) and the "natural" direction of rock 3.

Just another example of the funny physics of curling.

*The YouTube video at www.youtube.com/watch?v=xfVx1-emCV8 shows very clearly the unexpected direction taken by the back rock. Unexpected for most people, though in the video you see that the experienced Canadian skip Jennifer Jones knows what happens when rocks freeze together.

The causes of curling may be multifactorial. If that is the case, it may be some time before we understand what is going on. If, for example, there are two physical reasons for curling, we will need to understand both of them before we understand the whole thing. Very likely, friction is one of the contributors. Perhaps another contributor is the *impulse* idea. The impulse model of curling posits that curling results from the rock striking ice pebbles as it slides down the ice; each of these pebbles provides a small impulse or force that pushes against the rock. Such a model can lead to asymmetric forces, and the correct form of curling behavior, by incorporating the debris particles that are chipped off the pebbles as the rock passes over them. Such debris is carried around the leading edge of the running band, as suggested in figure 4.7. The distribution of debris is asymmetrical: there is naturally more debris on the left side of a counterclockwise rotating rock than on the right side. The debris may influence the impulse force just discussed, and so the total force acting upon the rock may be asymmetrical, generating curl in just the right way.[14] A lot more experimental and theoretical work will be needed before this or any other theory of curling is widely accepted as being close to the truth. What happens underneath a curling rock is still a mystery.

14. I must confess to being biased in favor of this impulse theory, being its author.

PART II

Snow Sports

5 SKIING—ON THE SLOPES AND ON THE LEVEL

There are many images from alpine skiing events at the 2010 Winter Olympics that stand out. You may recall Lindsey Vonn flying down the slopes at Whistler Creekside, bringing to that race a pedigree as world champion and a shin injury, and finishing the race with a gold medal—America's first in women's downhill. In the men's 50-km cross-country event, Norway's Petter Northug just beat Germany's Axel Teichmann in a *sprint finish* to this grueling competition. These races took place in the coastal mountains above the city of Vancouver. Snow sports, unlike ice sports at the Olympics, necessarily take place outdoors. As a consequence, both track and meteorological conditions are much more variable. Weather affects snow properties, which affect track condition. Altitude influences both air resistance and athlete performance (the finish lines at Whistler Creekside, where all the 2010 alpine skiing events took place, are 2,700 feet above sea level).

Despite this variability, we can still analyze alpine and cross-country skiing from a physics perspective and understand what is going on. The predictions we make will not be as accurate as for ice sports because of the unpredictability of conditions. For example, we must expect considerable differences in snow properties over a long-distance cross-country event, as the skiers pass from flatter to steeper terrain with humidity and shade variations, and variation in ambient temperature as the event progresses. Physics underlies everything, however, and in this chapter we apply it to alpine and cross-country skiing.

GETTING HIGH

Many outdoor winter sports, such as skiing, take place at high latitudes or high altitudes to ensure a good covering of snow. Altitude influences performance significantly, for better and for worse, even for indoor sports. For example, in the 2010 Winter Olympic Games in Vancouver, no world records were set in long-track speed skating. (Only two Olympic records were set: Sven Kramer of the Netherlands in the 5,000-m event and Lee Seung-Hoon of South Korea at the 10,000-m event.) The reason is very clear. The facilities for the 2010 Games were excellent, but the skating track (the Richmond Olympic Oval) was at sea level. Similarly, no records were set during the Turin Winter Games in 2006. Turin is less than 800 feet above sea level. All men's and women's long-track speed skating world records, at all distances, have been set at altitude—mostly at Calgary (altitude 3,400 ft) or Salt Lake City (altitude 4,300 ft).

The reason why altitude matters is, of course, because thin air at high altitude reduces aerodynamic drag. We saw in chapter 2 that for speed skaters, something like 80% of the power they expend during a race is used to overcome drag. Increased altitude reduces an athlete's oxygen supply during a race; this, you might think, would make fast racing more difficult at high-altitude venues, but drag wins out: speed skaters who want to set world records do much better when high (up). The same argument applies to other racing events, such as downhill skiing, where speed is important. The influence of altitude upon race times is investigated in technical note 10, where I show that, especially for short races, rarefied air leads to faster races.

The effects of altitude go beyond racing events, however. They influence all winter sports participants—not just downhill skiers or speed skaters. Athletes who normally compete near sea level have to make several adjustments when training for, and competing in, events at altitude. When training at altitude, they should increase the rest periods between exhaustive training sessions. It is a good idea for them, ahead of their competition event, to sleep and spend their free time at a higher altitude than the event venue. In competitions where it is permitted, athletes should supplement their oxygen intake on the sidelines or between events. In team sports such as hockey, substitutions should be made more often during games played at altitude.

It isn't just endurance that differs in the thin air of high-altitude competitions. Ski jumpers must adjust their lean angle, and there are similar technical adjustments necessary for downhill skiing and snowboarding. Athletes must take extra time to practice for the changes in speed and in projectile motion. After all, an athlete practice session involves thousands of coordinated muscle movements, and these patterns of movement become ingrained—second nature. They may need to be adjusted slightly in thin air because of the reduced air resistance and because the athlete's muscle response is different than at sea level.[*]

[*]An entire issue of the technical journal *Experimental Physiology* was dedicated to the physiological issues facing the athletes who competed at the Vancouver Winter Olympic Games in 2010. Many winter sports are played on mountains, of course, and so altitude issues are particularly relevant. See, e.g., Chapman, Stickford, and LeVine (2010).

SKIS

Skiing has a history that is at least as long as that of ice skating, and so it is no surprise to find that ski sports have diversified considerably over the years: there are aerial competitions, downhill and cross-country events, combinations of downhill skiing and aerial maneuvers, and combined cross-country and rifle shooting events. Given this multiplicity of sports, we can expect the same kind of diversity of ski equipment as we found for skating. Cross-country skis and ski poles are different from those used in alpine events, for example. Here we will look at the differences and see why and how they have developed.

Originally little more than wooden planks, skis have changed much in recent decades due to advances in technology. Modern skis may still have a wooden core (some have a foam core), but that is surrounded by a strong fiberglass box, with steel embedded at the edges. The base of modern skis is usually made from a polyethylene called P-Tex,[1] though graphite bases still can be found. Skis are *cambered*, which is to say they are bent so that they are higher in the middle than at the tip or tail, as illustrated in fig-

1. The type of polyethylene used for the base (underside) of skis goes by the catchy name of UHMWPE, which stands for ultra-high-molecular-weight polyethylene. This material is durable and slides easily over snow. It also absorbs ski wax very well.

ure 5.1. The stiffness of the ski and the weight of the skier determine how much of the camber is reduced when the skier stands upright on the skis. Cross-country skis are stiff so that, with weight evenly spread over both skis, the camber is still evident; alpine skis are generally less stiff. Indeed, as we will see, modern skis are designed to produce *reverse camber* during a turn—that is, the central part of the ski is lower than the tip or tail.[2]

Skis are fitted to individual skiers: they are as personalized as shoes or ice skates. Ski stiffness and length are adjusted to the skier's weight. Longer skis reduce pressure on snow and so are adopted by heavier skiers. Ski length also depends upon the skier's skill level. Experts opt for stiff, long skis, for speed and shock absorption. But since these skis are more difficult to turn, less-experienced skiers may prefer softer (less stiff), shorter skis. Such skis must be wider, if they are to have the same area, so that pressure on the snow does not increase.

Cross-country skis are narrow because the lessened width reduces sliding friction, which, as we will see, dominates the energy budget of cross-country skiers. As a consequence of being narrow, these skis must also be long. The *binding* (attachment to ski boots) is set further back in cross-country skis than in other types, so that the skier can run (actually the gait is a very awkward, energy-sapping waddle) up hill without tripping.[3] The binding in cross-country skis has a free heel, a bit like clap skates, so that the heel can lift away from the ski when the skier is poling over level ground or uphill. (Ski-jumpers also employ free-heel binding, so that they can lean forward during flight.)

Some alpine skis are twin-tip, which is to say turned up at the back as well as the front. This feature allows for skiing backward, and it facilitates landing from aerial maneuvers, as in mogul or freestyle competitions.

Modern skis differ from the classical skis of old in that they are *side-cut* to varying degrees. That is, the width of the ski is greater at the tip and tail than it is in the middle, near the binding, as sketched (in greatly

---

2. Manufacturers must walk a tightrope to get the ski stiffness just right. While stiffness must not be too great when applying downward force, it must be very large about the longitudinal axis of the ski so that the ski does not twist about its length when stressed. Such longitudinal twisting would adversely change the edge geometry.

3. Despite looking so awkward, the gait of a running skier is vaguely reminiscent of the way that skaters accelerate from a standing start because they both move by pushing sideways. Cross-country skiers are assisted by long ski poles, which change the dynamics, as we see in the text of this chapter.

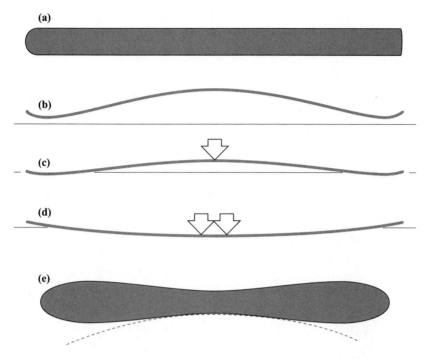

Figure 5.1. Ski geometry. (a) A classical straight-sided ski, seen from above. (b) From the side, the camber is clearly seen. (c) When a cross-country skier's weight is evenly distributed over both skis, the camber is still present, but not when (d) most of the weight is on one ski, as when ambulating. (e) The parabolic, or side-cut, ski. The degree of side-cutting is here greatly exaggerated. The dashed line is part of a circle.

exaggerated fashion) in figure 5.1. The shape of the ski edge on both sides is close to a circle of a radius between about 19 m and 24 m. This innovation is a recent development imported from snowboarding, appearing in the 1990s, and has come to dominate ski design, because it makes turning on skis much easier and faster. This revolutionary type of ski is known as *parabolic*, from the exact shape of the ski edges, for technical reasons connected with the way in which they assist with turning. This is explained in technical note 11 and in the next section. A parabolic ski is shown in figure 5.2.

There are many other differences between ski equipment used in the various ski disciplines, arising from skier characteristics as well as from the demands of particular sports. Women's skis tend to be softer, with the

Figure 5.2. A parabolic ski. These skis have the same overall area as classical straight-edged skis but are shorter because the parabolic ski is fatter at front and back.

binding further forward to increase stability and to facilitate turns. Alpine skis are not free-heel: the ski boot is bound to the ski at heel and toe. Slalom skis have a deeper side cut (i.e., a shorter turn radius) than those of downhill or super-G, because slalom skiing is less fast and has tighter turns. Cross-country ski poles are longer than those used in other ski competitions. Ski waxes differ with ski sport as well as with snow conditions.

Ski waxes. Here is a subject that arouses unbridled passions in the breasts of some skiers and is big enough for a book of its own—but it is not central to our explorations here, so I will not wax lyrical upon it.[4] Suffice it to say that there are many types of wax that can be applied to the base of skis to increase their slipperiness—or reduce their sliding friction— over snow. The type of wax depends upon the type of snow. Different waxes are applied to the ski base in different ways, and various waxes can be mixed together to form the best combination for particular snow conditions. The entire base area is treated with such *glide wax*.

For cross-country skiing, the skier wants skis that slide over snow when she is sliding downhill, but which grip the snow when she is struggling up a hill. Skis can be made both grippy and slippy by applying glide wax at the tip and tail of the ski, and *grip wax* in the middle, underneath the skier's feet (the *kick zone*). Grip waxes increase grip (duh), but only when the central part of the ski base touches the snow. Recall that ski camber keeps this part of the ski

4. An accessible account of ski waxes and their significance can be found in Karydas (2008).

off the snow when the skier's weight is evenly distributed, as when she is skiing downhill.[5] When she is plodding uphill, however, or poling across level ground, she applies force to one ski which presses the center down into the snow, thus bringing the grip wax into contact with the surface. Clever, hmm?

ONE GOOD TURN LEADS TO ANOTHER

Turning on skis is entirely different from turning on skates. The techniques for turning have evolved over the past century or so, as understanding of movement over snow grew and, most importantly, as the technology of skis developed. Turning on modern parabolic skis is not like turning on classical straight skis: all that effort your parents or grandparents put into learning to perfect those elegant parallel turns is now history.

Before getting into the mechanics of ski turns, let's consider the forces that act upon a skier. In figure 5.3 you can see our old friends gravity, sliding friction, and aerodynamic drag. There is a subtle difference, however, between the influence these forces have upon a skier and the influence they have upon a skater, or a bobsledder. Gravity acts at the skier's center of gravity (CG), drag acts at her center of pressure (usually pretty close to the CG), whereas sliding friction acts at her feet. This is different from bobsled or luge or skeleton, where all three forces acted at more or less the same place (because a sled's CG is close to the ground). The significance of the difference is that the combination of forces acting upon a skier will exert a torque, causing the skier to fall backward or tip forward unless she does something about it. For skaters, the problem is not so acute because sliding friction is much less important than it is for skiers (we will get to the numbers soon enough). A skier has to lean forward at an angle in order to counter the torque that acts upon her body.

This requirement has nothing to do with crouching in the tuck position to minimize drag: a skier standing straight would still have to lean forward to counter the torque. The lean angle depends upon speed. We could calculate it easily enough, but the details aren't important to us; all you need to know is that such leaning (*inclination*, in the technical jargon) is necessary. A consequence of inclination is the uneven distribution of weight upon the skis, and thus of pressure on the snow. That pressure was

5. This is why it is very important to match a skier's weight with ski stiffness. Too soft and the kick zone grips the snow surface when it shouldn't; too stiff and the kick zone never grips the snow.

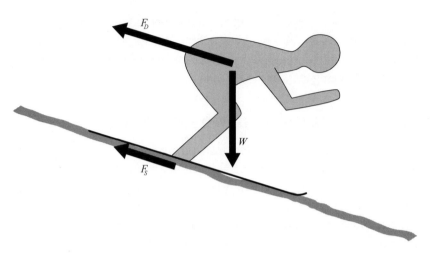

Figure 5.3. Force diagram for skiing. Note that the forces of gravity (W), sliding friction (F_S), and aerodynamic drag (F_D) act at different places; this induces torque.

already uneven because of ski camber. Both camber and inclination change dynamically as the skier maneuvers past gates or negotiates slopes, and this undoubtedly influences the sliding friction force (which depends upon the normal force, or force perpendicular to the surface, as you may recall from chapter 1). For our purposes, however, calculations will be sufficiently accurate if we ignore the uneven and rapidly changing pressure distribution beneath a ski and simply say that the friction force is a constant, as we have assumed in earlier chapters.

Skiers also *angulate* as well as incline. Angulation means that the skier's body is not straight; she adjusts her shape—bends her knees, crouches, leans to left or right—to best suit the circumstances she encounters. Correct (appropriate and timely) angulation is essential for turning on skis, be it old-fashioned stem or parallel turns or the new-fangled carving turns. The photo of figure 5.4 is a good illustration of angulation.

Beginners are taught the well-named *snowplow* or *wedge* turn, in which the skis are turned inward so that the tips are closer together than the tails. This action slows down the skier; by transferring weight to the left or right foot, the skier can then turn right or left. Snowplow turns are slow and work best on gentle slopes. The *stem turn* is a natural progression of the snowplow, for intermediate-level skiers. The outer ski (the ski that will be on the outside of the intended curve) is pointed in the new direction so

Figure 5.4. Lindsey Vonn angulates—distributes her weight—to take a gate at maximum speed. Photo by Thomas Vonn, courtesy of the U.S. Ski and Snowboard Association.

that, *a la* snowplow, the tips of the two skis are closer together. Thus, to execute a left turn, the right ski is pointed to the left; for a right turn, the left ski is pointed to the right. Then the skier transfers weight to the outer ski and points the other ski in the new direction. It sounds complicated, but the action is very simple. For big turns, several such stem maneuvers can be strung together.

Until the late 1980s, the method that most advanced skiers used for turning was the elegant *parallel turn*. The feet were brought close together, the skier leaned in the direction of the turn so that the edge of the skis bit into the snow surface. Weight was transferred to the outer ski. If performed correctly—it was a difficult maneuver—the turn would seem effortless and would look graceful, as it involved no skidding.[6] Consequently it was the fastest way to turn on skis. That is, competitive skiers would lose less time in a race by executing parallel turns than by using any other type. Actually, the parallel turn was a natural progression from the stem turn: it

6. Sometimes people being taught the parallel turn would be asked to hold something, such as a handkerchief, between their knees to encourage them to keep their feet together. The idea was that they were to ski down a slope, executing parallel turns, and arrive at the bottom with the handkerchief still in place.

gets its name from the fact that the skis are supposed to be parallel during the turn, but in fact they formed a slight wedge. A skier who evolves from beginner to intermediate to advanced executes snowplow turns, then stem, and then parallel. This progression requires movements from the skier that are less and less pigeon-toed. Because the wedge shape of the skis becomes less pronounced, there is less snow plowed aside and thus less slowing down during the turn maneuver.

The parallel turn is now relegated to history, thanks to the evolution of parabolic skis. The technique for turning on parabolic skis is known as *carving*, and it is easier and better than executing parallel turns. That is, correct execution of a carve turn is easier to master than correct execution of a parallel turn, and the turn can be faster. When executed properly, the ski carves a neat circular arc in the snow, with little or no snow thrown out sideways. Since the mid-1990s, carve turns have dominated skiing. The carve turn is driven by the side-cut geometry of the ski. Leaning into the curve and transferring weight to the outer ski causes the ski edge (the inside edge of the outer ski) to bite the snow. The extra weight forces down the middle of the ski, so that the camber is reversed—the ski bends up at tip and tail. The skier's legs are not brought together during a carve turn. The carve-turn action, and the shape of the side-cut, means that the ski edge is a circular arc, and so the ski executes a curve by carving a circle in the snow.[7] There is no slipping or skidding, and thus little or no snow is pushed aside.[8] Furthermore, there is no braking action for this turn. The result is a neat turn that does not slow down the skier very much. Skiing on the edges during a carve turn is shown in figure 5.5.

We can gain some insight into why the different types of turns are slow or fast by considering how much snow each type turns aside. Obviously it takes energy to move snow,[9] and this energy detracts from the skier's speed. More energy wasted in moving snow means less energy available for maintaining speed. You can see how this idea works from figure 5.6, which shows snow-

7. The fact that side-cut skis naturally produce turns has been described as their "yearn to turn"; see Pierce (1997).

8. If the inclination angle is larger than 30°–45°, the snow resistance to pressure decreases, and snow is pushed out of the indentation made by the ski. This can result in skidding and in large amounts of snow being sprayed out from the track, both of which absorb energy from the skier and reduce speed.

9. To put it technically, snow deforms inelastically, meaning that it absorbs energy and does not give it back. (Rubber deforms elastically: push it and it pushes back.) So the act of moving snow absorbs energy.

Figure 5.5. Erik Schlopy demonstrates a carve turn. His weight is on the outer foot; the skis are separated (unlike the old parallel turn) and are on edge. Only a small amount of snow is sprayed to one side. Photo by Gary Dickey, courtesy of the U.S. Ski and Snowboard Association.

plow, parallel, and carve turn. The snowplow disturbs more snow than the parallel, which disturbs more snow than the carve turn. The amount of snow shifted during a turn is not the whole story, of course—energy is also lost through increased friction coefficients and ski flexing, for example— but it is an indicator. Those dramatic-looking turns that cause snow to spray out like a lady's fan look cool, but they are slowing the skier down.

Another benefit of carve turns is that they sap less energy from the skier than did the older parallel turns. This is because the flex in the side-cut ski absorbs energy and gives it back when the skier links turns, jumping from, say, a left turn into a right turn. From an energy perspective, a skier who progresses down a hill executing left turns and then right turns to negotiate gates is *bouncing* down the hill, as if on a pogo stick. This is hyperbole— I am deliberately exaggerating to make a point—but the underlying idea is valid. Energy is stored in parabolic skis during a turn, some of which is returned to the skier when she emerges from the turn.[10]

10. Ski turns are discussed in, for example, Federolf (2006), Howe (1983), and Witherell (1988).

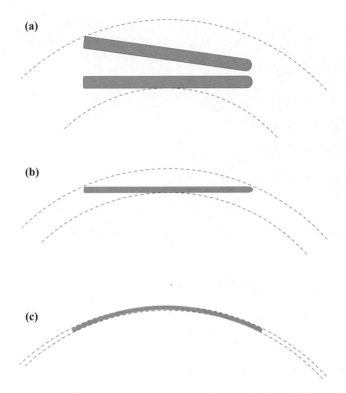

Figure 5.6. Ski turns. (a) A snowplow turn pushes aside a lot of snow. The area of snow affected is indicated by the dashed lines. (b) For a parallel turn, the outer ski presents an edge to the snow: weight is placed predominantly on this ski, and so the outer ski moves most of the snow. (c) A carve turn moves little snow: because of the side-cut geometry the ski edge carves a neat line. Clearly, the area of snow that is moved by this method is the least; thus, carve turns waste the least energy.

SKIING FRICTION AND DRAG

Three factors influence the speed of a skier in a race: the skier's weight, aerodynamic drag, and snow friction. We saw that for bobsled and the other sliding events, weight influences speed (and therefore time) because heavier sliders are less influenced by aerodynamic drag than are light sliders. The same physics applies to skiers descending a slope, though the influence of weight is less significant because aerodynamic drag is less important, as we are about to see. Basically, drag matters less because speeds are slower in skiing than in bobsled, luge, or skeleton. Sliding

Figure 5.7. American skier Jimmy Cochran adopts the tuck position to minimize aerodynamic drag during a U.S. alpine championship competition. Photo by Jen Desmond, courtesy of the U.S. Ski and Snowboard Association.

friction—here, snow friction—is more important, however; and this is why skiers go to so much effort to reduce the sliding friction coefficient as much as possible, by careful ski design, wax application, and skiing technique.

Skiers, unlike bobsledders, are not allowed to add ballast to increase their overall weight. Their drag coefficient can be reduced by wearing ski suits made of a low-drag material, and so we have the modern skin-tight, close-woven "slippery" suits (along with gloves and helmets that also serve to reduce drag). Manufacturers claim that form-hugging polyester suits reduce drag by several percent. Cross-sectional area—the body area presented to the airflow—is clearly something that the skier can influence. For example, skiers will adopt the tuck position (fig. 5.7) when possible, to reduce their size and thus reduce the aerodynamic drag force that acts upon them.[11]

Much research has been devoted to measuring and minimizing friction and drag. First, the measurements. Recall that aerodynamic drag force is

11. For skier cross-sectional area, upright and during a tuck, see Leino, Spring, and Suominen (1983).

characterized by a drag factor, b (see technical note 3). To minimize drag force, we need to reduce b to its smallest value. We saw that b increases with drag coefficient, air density, and skier cross-sectional area, and decreases with skier weight.

Putting together the data of different researchers we come up with a drag factor in the range $b = 0.0010$ m^{-1} to $b = 0.0014$ m^{-1}. This drag factor varies because skiers' weight, size, and skiing action vary. In general it is similar to, though a little larger than, the drag factor for skaters. The sliding friction coefficient has been measured to be between $\mu = 0.05$ and $\mu = 0.20$. Two points immediately strike us about these numbers. First, even the lowest friction coefficient for skiers is more than 10 times larger than the friction coefficient for skaters. The reasons are obvious: snow is not as slippery as ice, and skis are larger than ice skate blades. Second, the range of values for the friction coefficient is very large. This large range is due to the variability of snow conditions. Dry snow produces a low friction coefficient, whereas the friction of wet snow is much greater.[12]

These are the numbers we have to work with. We can check that they are reasonable by looking at results from downhill skiing competitions and applying what we have learned about the physics of downhill motion, subject to the forces of gravity, sliding friction, and drag. We calculated such trajectories in technical note 6 and have applied them to bobsled and other sliding sports. The same equations apply here, but with the appropriate choice for drag coefficient and friction coefficient. Consider, for example, the men's downhill competition at the 2010 Winter Olympics. This competition took place at the Whistler Creekside course, over a race distance of 3,158 m with a vertical drop of 870 m. The winner was Didier Defago of Switzerland, who posted a time of 1:54.01, giving him an average speed of 99 kph. This was a very close contest: the silver medalist (Aksel Lund Svindal of Norway) finished only 0.07 seconds behind, while bronze medalist Bode Miller of the U.S.A. finished only 0.09 seconds behind the winner. This means that Miller was less than 8 feet away from a gold medal and 2 feet away from silver—after hurtling downhill for nearly 2 miles.[13]

12. Figures for drag coefficient are from Nachbauer, Kaps, and Mössner (1992). See also Kaman (2001, 193); Kaps, Nachbauer, and Mössner (1996); and Watkins (1998, 198). For skier sliding friction coefficient, see, e.g., Armenti (1984).

13. These guys know how to hurtle: their average speed exceeded 98 kph (61 mph). The ladies were even faster (with an average speed of 101 kph), though the competitors were not

Now let's plug these numbers, plus our measured range of values for the aerodynamic drag factor b, into the formulas we derived in technical note 6—in particular, for those readers who are following the math in detail, equations (N6.6), (N6.8), and (N6.9). We find that the terminal speed for the Whistler run on that day was in the range 114–121 ms^{-1} (depending upon the drag factor). This is higher than the average speed of the fastest skier, of course, because skiers have to change course, and that slows them down. It is interesting to note that the medalists achieved around 85% of the theoretical maximum speed. We also find that the sliding friction coefficient for the Whistler track on that day was in the range 0.14–0.17. This is within the measured limits for the friction coefficient of skis on snow (0.05–0.20, recall) and so we are encouraged to believe that our physical model is reasonable.[14]

ENERGY AND POWER

Here is another physics calculation about the men's downhill competition at the Whistler Creekside course, which introduces us to the energetics of the event. In technical note 12 we show what happens to the energy of the skier as he moves down the slope. At the start he possesses a lot of gravitational potential energy—he is at the top of the slope. As he picks up speed he is gaining kinetic energy (the energy of motion) and at the same time spending energy in overcoming sliding friction and aerodynamic drag.[15] This friction and drag energy is wasted, from the skier's perspective; it causes snow to be pushed aside and air to be moved, and both snow and air to be heated up a little. What is remarkable about this energy calculation is that it shows just how much of the energy is used to overcome these dissipative forces, and how little is used to provide the skier with speed. Of the gravitational energy stored in the skier at the start of his race, a little

---

nearly as close: Lindsey Vonn was more than half a second ahead of fellow American Julia Mancuso, who was a second ahead of the bronze medalist, Elisabeth Goergl from Austria.

14. This short exercise does not prove that our physical model of downhill skiing is correct, but it does show that the model makes predictions that are compatible with the real world. If, for example, our calculations had shown that the friction coefficient must exceed 1, or must be unrealistically small, then we would have been forced to consider the model to be inadequate.

15. If there were no sliding friction or aerodynamic drag, the alpine skiers would have crossed the finish line at Whistler Creekside at speeds approaching 300 mph.

over 56% is spent in overcoming sliding friction, and 39% is spent in overcoming aerodynamic drag; only 4.4% of the energy available at the start of the race is utilized by the skier to increase his speed. This shows just why it is so important to wear the lowest-drag ski suit, wax the skis to obtain the least sliding friction, and tuck and carve as expertly as possible to reduce these dissipative forces, because they soak up over 95% of the available gravitational energy.

What about cross-country skiing? In this case there is not much potential energy available from gravity. The race is up and down over level and slightly undulating ground, with no significant net slope. Where do cross-country skiers get their energy from? From within their own bodies. Recall how they fall over the finish line at the end of a race: they have pushed themselves to and beyond exhaustion. The numbers are incredible. Here's how they work out.

We will consider the men's 50-km cross-country competition. Petter Northug of Norway won with a time of 2:05:35.5, which meant that he moved with an average speed of 23.9 kph—nearly 15 mph. We saw earlier that this was a close race with a sprint finish; in fact, Northug was just 0.3 seconds in front of the German silver medalist. Let's assume that Northug's long and thin cross-country skis gliding over the snow generated a low sliding friction coefficient of $\mu = 0.05$. If the net gradient of the course was zero—start line and finish line at about the same altitude—we can calculate the power that he must have expended in order to overcome the friction and drag that held him back every step of the way. He exerted 260 W to overcome sliding friction and 35 W to overcome aerodynamic drag. His total power output was thus 295 W, and he maintained this output for over 2 hours.

To put this figure into perspective, you or I would be breathless maintaining that power output for just 5 minutes (unless you happen to be an athlete). Let's say that you have been promised a medal if you can climb a 9,000-foot ladder (that's 1¾ miles) in 2 hours—about the same power expenditure as the men's 50-km cross-country podium skiers. And our calculation for cross-country skiing assumes a low value for sliding friction, recall. If the actual friction coefficient encountered in the 50-km race was greater, then the medalists' power output would also be proportionally greater. More: this calculation does not include the power that cross-country skiers must exert to move their arms and legs; recall the ungainly, flailing manner in which they struggle uphill (see fig. 5.8), each scrabbling

Figure 5.8. Cross-country skiers struggle uphill during a U.S. Olympic trial event. Photo by Tom Kelly, courtesy of the U.S. Ski and Snowboard Association.

over the surface like a demented octopus (sorry, guys, but so it seems to me). I have included only the external power consumption expended to overcome friction and drag, and not the internal power needed to generate body movements. There has got to be an easier way to win a medal.

Suppose that Northug on that day had not waxed his skis in quite the same way. Suppose that his choice or application of wax was such that the average sliding friction coefficient he experienced had been greater by just 0.1%. Let's say he generates the same power in this hypothetical race as he did in the real race. Then his finishing time would have been slower by 6.6 seconds. He would have been knocked off the top of the podium. Now you can see why so much effort is put into waxing skis and designing them so that they slide as smoothly as possible over the snow. One more time: the differences between hero and zero are miniscule.

In shorter cross-country events, sliding friction is still the dominant factor, though not quite so much as for the marathon 50-km race. The men's 15-km competition at Whistler was won by Switzerland's Dario Cologna with a time of 33:36.3. His average speed was thus 26.8 kph. As-

suming the same friction and drag parameters, Cologna generated 341 W
of power, of which 85% was used to overcome sliding friction (compared
with 88% for the longer event). These figures contrast with skating and
bobsled, where the higher speeds mean that drag is the more significant
dissipative force.

GETTING INTO YOUR STRIDE

Cross-country skiing involves poling across level surfaces or uphill, and
gliding downhill without using the poles. Such downhill stretches provide
a welcome breathing space, during which the grateful cross-country skier
will crouch in the tuck position, poles horizontal, as in alpine skiing. Most
of the time, however, these doughty athletes apply one of three gaits, each
with a well-defined use of leg stride and pole. The three gaits differ in the
coordination of pole and leg strides and in the synchronization of the
poles (the poles may both be planted in the snow at the same time or at
different times within the stride).

Without delving into the details, I investigate a simple example in technical note 13. There, you will find a calculation that predicts much of the
known behavior of poling (for example, that skier speed increases with the
cycle length—the distance through which the skier travels while a pole is
planted in the snow). The simplifying assumptions made in this calculation are necessary to keep the math from becoming too tangled. So long as
the simplifications are not so severe that they distort the physics, then such
an approach is valid. It permits us to see the essential elements of a physical system without getting confused by inessential details. Thus, for example, I assume that the snowfield being traversed by the skier is perfectly
level. Also, I assume that the leg action and pole action can be lumped
together into one force during a stride cycle.

The three real gaits are more complicated than this, of course, and if you
want to know the details, they can be found in the technical literature.[16]
Technical note 13 provides an appetizer to such an entrée. If you are not
that hungry, then here are the main results. Skiing speed increases with
increasing stride frequency and stride length (no surprise there). A less

16. Details of three cross-country skiing gaits can be found online at, for example,
http://biomekanikk.nih.no/xchandbook/ski3.html (section on ski skating techniques in
Smith 2002), and in Rusko (2003, chap. 2).

obvious prediction involves the skier's *duty factor*. This important parameter is the fraction of the time that a skier is applying force to the poles. The duty factor is independent of the frequency with which the poles are used; rather, it refers to the duration of pole use within a stride. In technical note 13 we see that the best strategy for the skier is to either use the poles sparingly (a low duty factor) or all the time (a high duty factor). All else being the same, middling values of the duty factor produce lower speed.

6 SKI JUMPING AND SNOWBOARDING— ON SNOW AND AIR

The physics of winter sports takes off in this chapter, as we follow skiers and snowboarders who leave the slopes and go airborne, under the action of physical forces. We will still encounter our old friends gravity and aerodynamic drag, of course, but we leave sliding friction on the ground and take up with a new force—aerodynamic lift.

In fact, we will need to consider lift only in the context of ski jumping. We begin with the alpine skiers of the last chapter, who find themselves flying—without much aerodynamic lift—as they careen over a sharp hill. Once they have taken off, how far will they travel before landing on the slope again? We will take our first look at snowboarders and see why they *must* get airborne in order to be competitive. I am not referring here to the aerial tricks that are the key component of, say, halfpipe events, but to snowboard cross. This is the snowboard equivalent of the alpine ski event, and aerial tricks do not figure (now that is almost a pun). The winner of a snowboard cross event is the first one across the finish line. We will see that the winner has to be good in the air.

Ski jumpers are a breed apart. The Olympic authorities were not much enamored with Eddie "The Eagle" Edwards, the inept ski jumper from Great Britain who stole the show at the 1988 Calgary games, despite finishing last in both his events. Ski jumping is a hugely technical event, as we will see, and Edwards was viewed by some people close to the sport as an embarrassing distraction. We will see why he fared so badly, and why more capable and better-equipped jumpers can jump the length of a football pitch.[1]

---

1. In fact, ski jumpers can achieve distances that are more than twice the length of a football pitch in the extreme version of ski jumping competition, the ski-flying events. Ski

ALPINE AIR

We saw in chapter 5 that to get onto the podium in the 2010 Winter Games, a downhill skier needed to average about 100 kph (about 60 mph). She would became airborne at several points during each run. This is not surprising, given the bumpy nature of the slopes that these top-flight (hah!) competitors careened down (see fig. 6.1).

With a few simple assumptions about the shape of the slope, we can calculate how far a downhill skier will fly between takeoff and landing. Clearly, this is an important consideration: the skier will want to have a pretty good idea of where she is going to put down *before* she takes off (for example, she will not want to land on the wrong side of a gate) because a left-right direction change midflight is not going to happen. In fact, calculations show that a downhill skier can jump 50–60 m, as we see in technical note 14. This is quite a jump, and it lasts about 2 seconds. The calculation includes the effects of drag, as it must at such high speed, and it assumes the quite modest slope shown in figure 6.2a. We assume that the skier takes off from a level part of the course and lands on a 20° incline. Not only must downhill skiers anticipate the landing point and negotiate landing without falling or losing too much speed, but they must also maintain their balance midair. Most of the crashes that occur in downhill events are the results of the skier losing her balance in the air or in a bad landing.

Another question arises, concerning the jumps that a skier undertakes. Consider figure 6.2b. Here we imagine a skier presented with a choice during a downhill or mogul competition: jump or don't jump? That is, the skier may choose one of two routes; the first route takes her down the slope she is already on, whereas the second route leads to a jump. Which route is the quicker?

It may be important to choose the quicker route if run time is a critical component of scoring. In most instances the skier will not have much choice—the only available route will or will not lead to a jump—but sometimes there will be a choice. In mogul events, particularly, there are many small bumps, and the skier may have a choice of route (fig. 6.3).

------

flying is not part of the Olympic Games (there are only five accredited ski-flying hills in the world, all of them in Europe). The current world record distance for a ski-flying leap is 239 m (784 ft), achieved by Bjørn Einar Romøren of Norway in 2005. You can see this amazing jump on YouTube, at www.youtube.com/watch?v=NqKuKTBsQJk&feature= related.

Figure 6.1. Alpine skier Steve Nyman takes to the air during a downhill event in 2008. Photo by Gary Dickey, courtesy of the U.S. Ski and Snowboard Association.

Often, this choice will be made for reasons other than speed—perhaps the skier wants to position herself just right for one of the two major aerials that are required in mogul, for example. Nevertheless, time is a factor and so it is important to know which route of figure 6.2b is the faster.

The answer is not obvious. On the one hand, going airborne gets rid of sliding friction, and so we might expect this route to be faster. On the other hand, sliding over the level section just prior to takeoff will slow the skier down (compared with sliding down the slope). So which wins? The physics is laid out for you in technical note 15, and the results of this analysis are shown in figure 6.4. For the particular case of a skier moving with an initial speed of 54 kph (15 ms^{-1}), I have plotted in figure 6.4a the time difference (between sliding on the level and then jumping, or sliding continuously over the same distance) against slope angle. For gentle slopes there is a slight advantage in avoiding the jump: sliding is slightly faster. For steeper slopes there is an increasing advantage in taking the jump.

The calculation results are presented in a different way in figure 6.4b, where skier speed is plotted against slope angle. Sliding is faster in the lower left region of the graph, and jumping is faster in the upper right. These regions are separated by a line, shown on the graph, that corresponds to zero time difference, when jumping and sliding result in the same distance

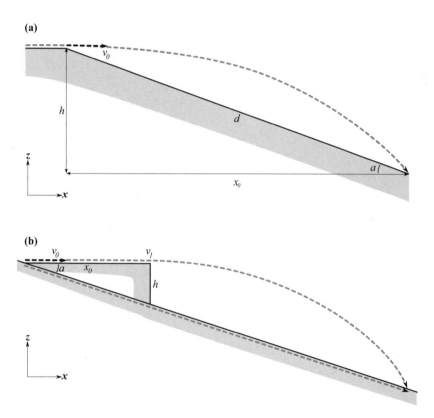

(a)

(b)

Figure 6.2. Jumping downhill. (a) A skier starts on the level with an initial speed of, say, $v_0 = 100$ kph, typical for downhill events. Say she takes off from a level section and lands on a $20°$ slope (so that the angle a is $20°$). How far does she fly? (b) Now our skier is faced with a choice of two routes, one of which involves a jump. Which route is faster?

Figure 6.3. Hannah Kearney at the 2010 Freestyle World Cup, in Calgary. Photo by Garth Hagar, courtesy of the U.S. Ski and Snowboard Association.

being covered in the same time. The position of the dividing line depends upon the length of the level section that precedes the jump (the distance x_0 of fig. 6.2b). You can see that shorter levels favor jumping, whereas longer levels do not.

FROM SNURFING TO SNOWBOARD SUCCESS

The snowboard is an American invention of the 1960s. Originally called the *snurfer* (for "snow surfer"), it has evolved into the modern high-tech snowboard over the intervening decades. A heady combination of surfing,

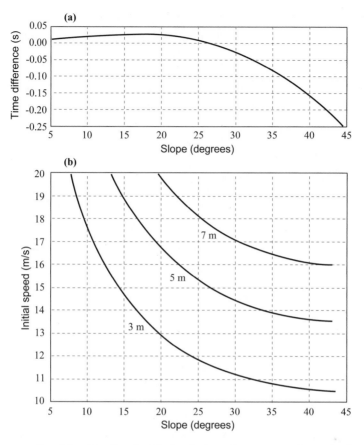

Figure 6.4. Should a skier faced with the choice of two routes shown in figure 6.2b jump or slide? (a) The time difference (jump minus slide) is plotted as a function of slope angle. Positive values mean that sliding is faster; negative values mean jumping is faster. In this case, the initial speed of the skier was chosen to be 15 ms^{-1}. (b) Initial speed vs. slope angle for different lengths of the level section (x_0 of fig. 6.2b). The lines correspond to zero time difference; jumping is faster for a combination of speed and slope that lies above the lines. The line labels refer to the length of level section x_0.

skiing, and skateboarding, snowboarding exploded as a hobby and then, inevitably, emerged as a major winter sport which has grown and grown and grown and grown and grown and grown. (I'm trying to emphasize here that the sport has grown.) The first world championship took place in 1985, at Livigno in Italy and St. Moritz in Switzerland. Snowboarding was adopted by the FIS (the International Ski Federation, the body that

Figure 6.5. Snowboarding a halfpipe. You can see from this picture that the board is side-cut, that it is attached to the snowboarder, and that it is often in the air. Aerial tricks (such as a *grab*, a *flip*, or a *spin* or a combination of these elements) are major components of halfpipe events, which take place on a semicylindrical surface like that shown here. Usually the surface is placed on a slope, so that the snowboarder progresses downhill as well as across the surface. Photo courtesy of the U.S. Ski and Snowboard Association.

oversees ski competition) in 1994. Four years later snowboarding appeared for the first time in the Winter Olympic Games, in Nagano, Japan. At the next Winter Olympics, in Salt Lake City, the number of snowboard competitions increased.

Today, around the world, snowboarding accounts for 40% of all sliding snow activities. Its popularity is acknowledged by almost all the major ski facilities, which provide specialized snowboarding surfaces such as the *halfpipe* (fig. 6.5). Initially, the skiing establishment looked upon this upstart new sport with some wariness, but today the popularity of snowboarding (and the revenue it generates) has overcome these misgivings.

You can see from figure 6.5 that the snowboard is a short, very wide ski that is attached to the snowboarder's feet at an angle. The board is waxed and side-cut like skis. The bindings do not release automatically in a fall, as happens for skis, and this has perhaps contributed to the different kind of injuries that result from the two sports. Most skiing injuries involve the knees, largely because of the way that skiers turn, which places great stress upon the knee joints. For snowboarders, the most common injuries involve wrists or the head. It is pretty clear from figure 6.5 why head injuries might occur: airborne tricks are a key aspect of halfpipe snowboarding, and a fall can be dangerous. Therefore, a helmet is mandatory in international competitions.

Snowboarders turn their boards by increasing pressure on the edges, as do skiers, but the skills required are different. The stance of the snowboarder—across the longitudinal axis instead of parallel to it—ensures that this is the case. Balance is very important for snowboarders. There are no poles to restore equilibrium or to generate speed. Snowboarding today has matured as a sport to the extent that it has specialized. The equipment—boots and boards—used for alpine events are different from those used for halfpipe or for snowboard-cross competition. The surfaces are very different in the different types of competition. Giant slalom events take place on a course about 550 m long with an average slope of 12°–20°; snowboarders negotiate gates (between 18 and 25 of them) while descending at speed.[2] The snowboard-cross course is longer (between 500 m and 900 m); competitions involve four snowboarders negotiating moguls, jumps, and sharp turns as fast as possible (fig. 6.6).

Snowboarding is generally more up-in-the-air than skiing. (I hope that this fact is effectively conveyed in figs. 6.5 and 6.6.) The boards are attached to the boarders for this very reason. Now it is time for me to turn to an analysis of this popular sport, to see just why snowboarders are so ready to head skyward. The reason is not simply to do tricks; speed events also lead to airborne trajectories. I will concentrate here on two examples.

JUMP TO IT

Look at the snowboarders of figure 6.7. One of them slides over a stretch of level ground, while another crouches and then jumps. They start out at the same place, on the left of the page, moving at the same speed. Which

2. The highest recorded speed for a snowboarder—a shade over 200 kph—was set by Australian Darren Powell back in 1999, in Les Arcs, France. This is quite a clip, but skiers are speedier: the speed record on skis is just over 250 kph. Now that you are an expert on sliding friction and aerodynamic drag, you can probably guess why skis are faster.

approach gets our snowboarders across the page faster? This situation is very similar to that of the downhill skiers we looked at in chapter 5, and the analysis is similar, so I won't repeat it. Here's the result. If the snowboarder who jumps reaches a height h, then she lands a little behind the snowboarder who opted to slide across the page. In fact, she lands a distance μh behind, where μ is the sliding friction coefficient. This distance is small—maybe only a few centimeters. But before you jump to the conclusion (so to speak) that jumping is not a good idea in this situation, consider the *speed* of the snowboarders at the time that the jumper lands. It turns out that the jumper is moving faster than the slider and will, a short time later, overtake the slider. So jumping may be worth the effort after all.

Why does this behavior occur? First, let's see why, when she lands, the jumper is behind the slider. During the acceleration phase of the jump, when the jumper is pushing down on the board in order to get airborne, she is effectively increasing her weight (increasing the normal force). As we saw in chapter 1, increased weight means that during this phase she

Figure 6.6. Four snowboarders compete at the 2010 Grand Prix at Park City Utah. Photo courtesy of the U.S. Ski and Snowboard Association.

Figure 6.7. Snowboarder options. Is it quicker for her to (a) slide or (b) jump? Assume a level surface.

experiences greater sliding friction. So, during takeoff, the jumper's forward progress is slowed. When she is airborne, however, most of this loss is offset by the fact that she is now subjected to zero sliding friction. (Of course, both snowboarders are subjected to drag force, but only the slider is held back by sliding friction while the jumper is airborne.) The offset is not quite complete, and the jumper lands a few centimeters behind the slider. What about the speeds of the two snowboarders? During takeoff, the jumper's forward speed slows, as we have seen. During the time she is airborne, however, the other boarder—the slider—slows down more. In this case, the jumper wins. Putting both results together, the jumper lands a little behind the slider, but because she is moving faster, she soon overtakes the slider.

So here is one reason why snowboarders jump a lot, even when they don't have to. For the level-surface situation just analyzed, the gains are probably quite small and may or may not be worth the effort. There is one circumstance when jumping most certainly *is* worth the effort, however, and this occurs at the start of some snowboard-cross events. Let me set up the scene for you, and then show why jumping is beneficial.

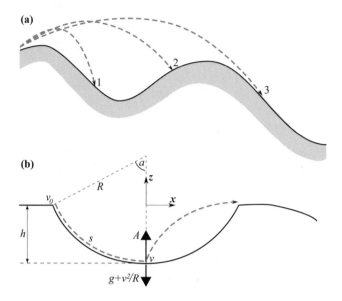

Figure 6.8. Snowboard-cross. (a) Jumping the hills must be finely judged. Trajectory 1 is safe but slow, trajectory 2 will probably result in a crash, and trajectory 3 is best: soft landing and high speed. (b) Slide or jump? Is it better for a snowboarder to slide through this trough, or to jump (as indicated by the dashed line)? In theory as well as practice, it is better to jump.

Snowboard-cross events usually consist of four competitors, who emerge from their start gates at the top of the run and then speed down the course, which consists of sharp bends and a variety of hills and jumps. In figure 6.8a you can see one of the challenges that face snowboarders as they descend the track. Their speed may be enough to get them airborne, in which case they have to judge carefully where they are going to land. Clearly, trajectory 2 of figure 6.8a is disastrous; such a hard landing may well result in a crash. Indeed, it is rare in Olympic snowboard-cross competitions for one snow-boarder to overtake another *except* when a competitor crashes.

Given this fact—that it is difficult to overtake a competitor once she has the lead (unless she makes a mistake)—the start becomes very important. You must get off to a flying start so that you find yourself in the lead. The start of some snowboard-cross tracks consists of a trough that is almost as high on the far side as it is at the start; a simple example is illustrated in figure 6.8b. Let us say that the start gates are just 1 m to the left of the trough shown in the figure; right from the get-go our group of four snowboarders find themselves in a situation which requires the decision considered earlier: do I jump or not? The choice is to slide through the initial trough or to jump it as shown in figure 6.8b. Most choose to jump, and we can show why this is the right choice.

The trough of figure 6.8b, right at the start of a snowboard-cross race, is quite small—less than waist-deep—and so it is quite feasible to jump out of it, as shown in the figure. Technique and timing are crucial, however. Get it wrong and you end up stranded in the trough, or sliding backward into it, having generated insufficient speed to get up the slope. In technical note 16 we look at the following case. Assume that if the snowboarder chooses to slide through the trough, she has just enough initial speed to get her to the other side. That is, her speed entering the trough will be slowed by friction to the extent that she gets out of the other side with zero speed. Suppose that her competitors start out with the same initial speed but they all decide to jump—let's say at the lowest point of the trough. Of course, they must judge the jump just right: they must have the height, and the forward speed at takeoff, to land out of the trough. The calculation of technical note 16 shows that the jumpers land with forward speed, and so with an advantage over the slider.

There is one condition that must be satisfied before jumping the trough of figure 6.8b is worthwhile, and that refers to its size and shape. The trough must be steep-sided and not too deep. We can regard this type of design for snowboard-cross start as being chosen to force the competitors to make a decision (jump!) and to exercise the utmost skill in executing it. In the snowboard-cross competition at the 2010 Winter Olympics, Canada's Maelle Ricker won the gold medal right at the start of the race. She was fastest out of the initial trough and kept her lead right to the end. The trough was too deep to jump from the bottom, as in our calculation, so her jump occurred higher up the far side, but you can see it quite clearly in video footage of the race.[3]

3. The start of the snowboard-cross track at Whistler was not the simple trough of figure 6.8, and it required more complicated movement than a single well-timed jump for Ricker to emerge from it in the lead. Canadian viewers can see from the YouTube video of the final race (www.youtube.com/watch?v=qcK3b01rNpk&playnext_from=TL&videos=f4kiMp 25P-Y&feature=rec-LGOUT-real_rev-rn-1r-2-HM) that she came out of the gate faster than her opponents, but also that she controlled her weight with at least one jump inside the trough. The simplified scenario of figure 6.8 captures the essential feature of this type of start: the height of the trough is such that snowboarders will not make it out of the other side with any speed at all if they simply slide.

THERE BE GOLD IN THEM THAR HILLS

We have genteelly partaken of appetizers by way of light refreshment; now we get to snarf the main course. That is to say, in the rest of this chapter we will come to grips with ski jumping as an event, as opposed to jumping while on skis. Ski jumping separates the men from the boys—and from the girls, according to the Olympic organizers, who (to date) have banned women from the competition. In fact, plenty of ladies seem to be just as hell-bent on hurtling through cold air as do the men. Some might say that ski jumping does not separate men from boys so much as separating sportsmen from lunatics. I won't go there; my task here is to explain to you the interesting physics that is so much a part of this bizarre event.

The rules of ski jumping, the equipment used for ski jumping, the techniques employed during ski jumping—all of these are governed by physics. Even the shape of the ski-jumping hill is determined by physics. Obtaining a gold medal in ski jumping is thus as much a matter of understanding the physics of the sport as it is of mastering the techniques of takeoff, flight, and landing.

The Hill

Not just any old hill will do; if a ski jumper wants to survive a bad landing, then, given the speed he is traveling when he hits the surface, he had better be traveling almost parallel to it. You can see how this consideration influences the shape of a ski-jumping hill. Before getting to that, let me start at the beginning. In figure 6.9 I have sketched the five phases of a ski jump. The *inrun* is the downhill section along which the jumper slides, building up speed. The *takeoff* is, I hope, self-evident. The *free-flight* phase ends with a *landing* and then a finish on a gently sloped *outrun* area. At the top of the hill there is a *scaffold*, as it is called—a tower with the start of the inrun at the top. Note that the drop-off, at the end of the inrun, is only about 3 m (10 ft). During the free-flight phase, the skier's trajectory is determined by the laws of gravity and aerodynamics. The shape of every part of the hill is specified by the governing body—the FIS—so that this trajectory is never very far from the surface of the hill. That is to say, the hill falls away beneath the skier almost as fast as the skier falls. Such is the theory, at least. It means that the jumper lands softly even while landing at high speed—more details below.

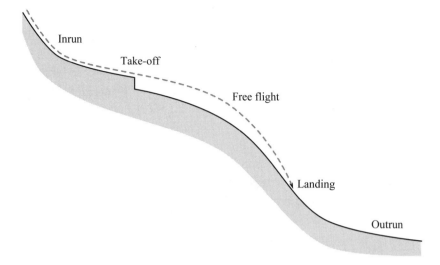

Figure 6.9. The five phases of a ski jump.

Competitive jumpers are expected to attain a par distance for a given inrun height. This is called the *K-point* (the "calculation point") and corresponds to a jump distance of 80–100 m for a *normal hill* and 120–140 m for a *large hill*.[4] Suppose a ski jumper does extraordinarily well and jumps way beyond the K-point. You can see from figure 6.9 that doing so makes the landing much more difficult because the hill slope becomes more gradual; that means that the skier lands at a much steeper angle and therefore hits the ground much harder.

The Sport

In Olympic competition, ski jumpers are awarded points for style as well as for distance. Each jumper has three jumps; the first one is for practice, and the last two both score. The scores are added, and the jumper with the most points wins the competition. The points are split about evenly between distance and style. Here's how. A jumper who lands at the K-point is awarded 60 points. If he exceeds the K-point he gets extra points using a formula that depends upon the hill. For the competition at Whistler during the Vancouver 2010 Olympics, the skier gained 1.8 points for every

4. Ski jumps from normal and large hills are part of Olympic competition. Ski flying is not; for ski flying the K-point is at least 185 m away from the takeoff point.

meter he jumped beyond the K-point. Similarly, he would be deducted 1.8 points for every meter he fell short of the K-point (perhaps I should say "landed short," because falling doesn't work in this sport).

There are five judges, positioned down near the K-point, who each award points out of 20 for style, based upon the technique displayed by the jumper during the inrun, takeoff, flight, and landing stages. For example, the skier is expected to land in the *Telemark* style, with one foot in front of the other. He is expected to be almost motionless during the flight. The five judges add up their scores, and the highest and lowest are discounted. The scores from the remaining three judges are added together to give the style points, and this sum is then added to the distance points to give the total points awarded for the jump.[5]

You would be forgiven for thinking that takeoff speed was a more important factor than style for determining finishing position. In fact, this might not be quite right. We will see in the next section that jump distance increases as takeoff speed increases, as you might expect, and so it would seem that finishing position would be improved by increasing takeoff speed, but it doesn't work like that. In practice, at the highest level of competition, jump distance is not strongly correlated with takeoff speed, as you can see for the Vancouver 2010 large-hill event in figure 6.10a. The reason for this surprising observation is that all the top jumpers seem to attain more or less the same initial speed. They all slide down the same inrun, and, as you can see from the graph, the variation in their initial speeds is only about 2%. The same graph shows that technique matters: note how the distances for the top 30 are longer than those of other competitors, even though the initial speeds were similar. This difference can only be due to technique during takeoff and flight.

In figure 6.10b, I plot, for each of the top 30 jumpers, the points the skier was awarded by judges against his finish position. Now there is a clear correlation: generally speaking, more style points resulted in a higher finish position. This is perhaps unsurprising, given that style accounts for about half of the total points. The winner, however—Simon Ammann of Switzerland—was not among the top 10 on style points. He won because of

5. Points can also be adjusted for changing weather conditions. Weather can greatly influence style points and distance (think of a jumper in midair, struggling to battle a crosswind); thus, changes in weather conditions during the course of a competition may unfairly disadvantage some of the competitors. The points adjustment is an attempt to counter this effect.

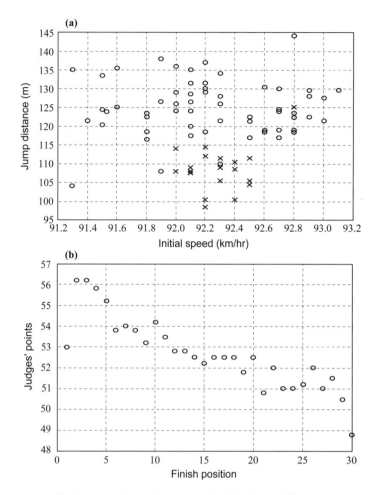

Figure 6.10. Ski-jump statistics from the individual large-hill competition at the 2010 Vancouver Olympics. (a) Jump distance vs. take-off speed, for the top-30 finishers (o) and for the other qualifiers (x). (b) Judges' points awarded for style vs. finishing position.

two massive jumps, one of which shattered the old Olympic record. So in this case, distance mattered more than style.[6]

Why should style count at all? After all, sports fans want to see long jump distances, not some artificial measurement of performance such as

6. Ammann's first jump of 144 m was 3 m longer than the previous Olympic record. The K-point of the large hill at Whistler is at 140 m, so Ammann earned an extra 7.2 points for distance. From figure 6.10b you can see that these extra points made up for his relative lack of style.

style. In fact, the two may go together, Ammann's performance notwith-standing. Good style may lead to longer jump distances by encouraging jumpers to adopt good technique, in the form of the most suitable pos-tures. Or it may not. We will see soon that the biggest single innovation introduced into the sport over the past 20 years has been the V-style of configuring the skis during flight. (Previously, skis were held parallel dur-ing flight.) At first, judges deducted style points for jumpers who adopted this innovation, as it was considered to look less elegant than the earlier posture.

Technique

A ski jumper ascends the tower by elevator. He sits at the top on a seat that straddles the inrun. When it is time to begin his run, he eases himself off the seat and drops into the "egg" position—body crouched and arms held back. (Until the 1970s, they were held forward.) This reduces his aero-dynamic drag. Of course, his suit is tight fitting and smooth, as for all high-speed winter sports. His skis are long—as long as those of cross-country skiers, though twice as wide. They have no metal edges because there are no turns in this sport. The binding is free-heel, and set back about 60% of the distance from the ski tip.

At the end of the inrun, at just the right time, the skier jumps, adding vertical speed to improve his distance. The takeoff angle of the ski jump is about 11° below horizontal. This is much less steep than the inrun slope and gives the false impression, when viewed from the bottom of the jump hill, that the skier is initially rising through the air. As soon as he is air-borne, the jumper quickly changes stance, moving from the crouched egg position to a straight-legged, rigid posture that minimizes aerodynamic drag and maximizes aerodynamic lift—lift is a critical component of this sport. Figure 6.11 shows a jumper in profile, adopting the right posture. There are many different parameters that are used to describe this posture, and even small changes to one of them can greatly influence the jump distance, by altering the lift and drag forces. Wind tunnel tests, of which there have been many over the last 20 years, show that 60% of the aero-dynamic forces act on the body of the jumper and 40% on the skis them-selves. The critical parameters are the ski *angle of attack* (the angle between the velocity direction of the jumper and the direction in which his skis point), the leg-ski angle, the leg-torso angle, and the ski V-angle.

Figure 6.11. Nick Alexander at the 2008 U.S. Ski Jump Championship. Photo by Tom Kelly, courtesy of the U.S. Ski and Snowboard Association.

I will get to the important V-angle in the next section; here let me say something about the other three parameters. In figure 6.11 you can see that the ski angle of attack is about 30°–40°, which is typical for ski jumping and corresponds to the values with best wind tunnel results. That is, ski angles of attack around 35° seem to produce the longest jump distances. Later I will manfully attempt to provide you with a mathematical analysis of ski-jump physics, but it will involve a number of simplifications to make the calculations tractable. The real world is more complicated, mostly because of the complex shape of the ski jumper. In fact, the most detailed mathematical analysis (described by horrendously complicated fluid-dynamical equations which can only be solved by a serious amount of computer number-crunching) is not particularly accurate. This is why researchers resort to the wind tunnel: it reproduces (approximately) the behavior of an airborne skier and permits the measurement of lift and drag forces.

You can see from figure 6.11 why the ski jumper is such a complicated projectile to describe. He can alter his ski angle of attack, as well as the ski-leg and leg-torso angles, quite significantly. He represents a projectile with many *degrees of freedom*, to use a mathematical term; he is not simply a point particle or a symmetrical cannonball. He is able, during flight, to influence the trajectory quite strongly by adjusting his configuration. I will reserve further comment on the lift and drag forces that act upon a ski jumper until the next section. For now, please just note that these forces are variable, hard to analyze, and partially controllable by the jumper.

So now our jumper lands, Telemark style, and slows to a halt amid cheering fans, if all went well. The speed with which he hits the slope of the ski hill is quite low, because the shape of the ski hill is chosen to match very nearly to the expected skier trajectory: he is never very far off the ground. In fact, observations show that the force felt by the skier upon landing near the K-point is no more than he would feel if he were dropped from a height of 0.5 m; for such a short drop, landing speed would be very low. This shows how well the ski-hill shape is matched to the ski jumper's trajectory because the actual speed at which he lands is 50 times bigger— but almost all of the velocity is parallel to the surface. At larger jump distances, beyond the K-point, the landing force increases rapidly, to the force he would experience if he were dropped from a height of 3 m (10 ft).[7]

LIFT, DRAG, AND V-STYLE

One aspect of the flight of ski jumpers is not apparent in the side view shown in figure 6.11, and that is the much-discussed V-angle of the skis. This ski configuration, shown in figure 6.12, represents the biggest single change in ski jumping over the last quarter century.

Before 1985, ski jumpers kept their skis parallel, and together, during flight. Indeed, as we have seen, style points were deducted from their performance if they did not do so. In that year, Jan Boklöv of Sweden adopted the now-standard V-style.[8] It is claimed to increase jump distance by 10% compared with parallel skis, as a result of increasing aerodynamic

7. The aerodynamic forces that act upon a ski jumper have been measured by, e.g., Seo, Watanabe, and Murikami (2004). The relative influence of torso and skis upon these forces, and the influence of ski suit design, is given by Luhtanen, Kivekäs, and Pulli (2000). Müller and Schmölzer (2002) discuss the shape of the ski hill and the force experienced by the jumper upon landing. A popular account of the physics of ski jumping can be seen in the YouTube video www.youtube.com/watch?v=BDpxSLv89Y8.

8. Boklöv certainly made the V-style well known, but he may not have been the first, or the only, skier to use this technique. Some sources claim that a Polish jumper, Miroslaw Graf, had been making use of this technique in the 1970s. See, for example, Maryniak, Ladyzynska-Kozdras, and Tomczak (2009).

Figure 6.12. Brett Camerota flies, at the Utah Olympic Park, Normal Hill, in August 2000. Note the V-angle of his skis. Photo by Marvin Kimble, courtesy of the U.S. Ski and Snowboard Association.

lift by 30%. I don't know about these figures, but the basic idea is not in doubt: configuring the skis in a V (with the angle at about 30°) increases jump distance because it increases lift force. (Interestingly, however, the world record distance for a ski jump, that of Bjørn Einar Romøren in 2005—see note 1 to this chapter—was achieved using a much more open style, with the skis more nearly parallel than V-shaped, but spread out wide, not held together as in the old days.)

We have encountered aerodynamic drag before in this book, several times; it is a pretty intuitive concept and easy to get a handle on. However its partner, aerodynamic lift, is another story. Lift is often misunderstood, and yet these two forces (lift and drag) go together like Laurel and Hardy. They are two sides of the same coin. Both act upon a moving object, and they depend upon the object speed in the same way. Both are a consequence of the fluid dynamics that applies to the air as our object flies through it. There are two key differences between lift and drag forces, however. First, they depend upon the shape of the moving object in different ways, and second (as a consequence) the lift force acts in a different direction. While drag always acts in the opposite direction to object velocity (hence the descriptive name, "drag"), lift acts perpendicular to the velocity.

"Hmm," you say, "but there is only one direction that opposes the velocity, whereas there are many directions that are perpendicular to it. For example, if our flying object (say an airplane) is heading due north on a level flight path, then east, west, up and down are all perpendicular to it. So which of these applies to the lift force?" This is where the object's shape comes into the mix. An airplane wing is shaped to create lift in the up direction. (I hope this is not too shocking a revelation to you.) An important factor in determining the lift force of a flying object is the angle of attack that the object (here, an airplane's wings) present to the airflow. The theory underlying lift—explaining how much force you get for a given airfoil shape presented at a given angle of attack—is complicated, and I will not go there. Please just accept that certain jumper configurations— such as V-style—generate more upward lift than other configurations. Jumpers found this out empirically, by trial and error; and now physicists are trying to catch up, to explain *why* this configuration is better.[9]

The lift of a ski jumper is not really like the lift of an airplane wing; apart from the very different shape of the two objects, they operate at very different angles of attack. An airplane wing might employ an angle of attack of, say, 4°, whereas our ski jumper's angle of attack is 10 times larger. He more closely resembles a flying plate, or a discus, than a wing. Anyway, his angle of attack changes throughout his flight—indeed, calculations suggest that the longest jump distances occur for jumpers who increase their angle of attack as their flight progresses. A small initial angle of attack serves to reduce drag, and a later large angle of attack increases lift. The dependence of lift and drag forces upon angle of attack has been measured, and the graphs are shown in figure 6.13. The details change as other

9. Readers who want to pursue this matter further and gain a deeper understanding of aerodynamic lift without doing a Ph.D. in aeronautics might consult the appendix of my sailing book (Denny 2009). For a practical demonstration, hold your hand flat, out of a moving car window. (Ahem, please do this as a passenger, not as a driver.) Adjust the angle of your palm to the airflow, and you will soon see how the lift force changes with angle of attack.

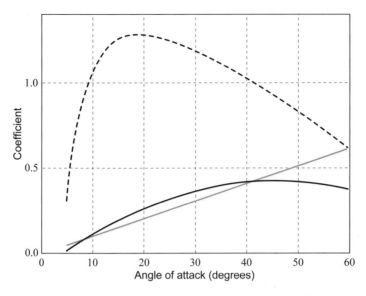

Figure 6.13. Aerodynamic force coefficients vs. angle of attack. Plotted are drag coefficient (gray line), lift coefficient (solid black line), and lift-to-drag ratio (dashed line). Data from de Mestre (1990).

parameters change, but the basic shape is usually as shown in the figure: drag increases pretty much proportionally with angle of attack, whereas lift increases and then falls away. An important parameter is the ratio of lift to drag force, which has the characteristic humpback shape of figure 6.13.

CALCULATING THE TRAJECTORY

A detailed mathematical analysis of ski jumping physics is not going to reduce the national deficit or cure the common cold. In fact, it isn't even going to describe accurately the trajectory of a ski jumper. Yet when we applied the same type of analysis to bobsledding, for example, we obtained results that were pretty darn close to the real thing. Why the difference? We've just seen why: the lift and drag change with angle of attack, which changes throughout the trajectory. Also, the jumper can adjust his configuration as he flies like an eagle (or like a turkey, as we will soon see) so that the lift and drag forces change again.

We have seen that detailed theoretical calculations are not really doing it for us, and so researchers have turned to wind tunnel tests. These are not ideal because they do not faithfully reproduce the conditions that a jumper experiences when in flight, but they are pretty good and have taught us a lot, in conjunction with detailed observations of many ski jumps. For example, we have learned that the crucial parameters that determine how far a jumper flies are his initial velocity, his flight configuration, and his stability in flight. Wind blowing up the hill will increase his distance; wind blowing down the hill will shorten it. Low jumper weight helps to increase flight distance. The lift-to-drag ratio can be maintained at a fairly constant and high level (exceeding 1) during flight.[10]

Can we make sense of these results, and can we get close enough to them mathematically with a simple analysis? Let's begin with the crucial parameters. Figure 6.10a shows that, at the top level, takeoff speed (the magnitude of takeoff velocity) doesn't influence jump distance much, for the simple reason that most of the Olympics jumpers achieve the same takeoff speed, more or less. But looking over a broader range of ski jumpers, of widely differing skill levels, reveals a different story. A lower skill level produces a much lower takeoff speed, and this indeed reduces jump distance, as we would intuitively expect. The vertical component of takeoff velocity (here I will call it the *initial lift speed*) is important because it determines the angle at which the jumper leaves the inrun. Recall that inruns end at a downward angle of about 11°; a skier who jumps upward at the last instant will reduce this angle and so will begin his flight in a more nearly horizontal direction. This action increases his jump distance.

Flight configuration and stability reflect the lift and drag forces that act upon a jumper. If he is able to maintain a high lift-to-drag ratio, he will go far in this world—literally. Wind speed increases or decreases lift force, and correspondingly influences jump distance. All of these effects are understood, approximately if not exactly, and a simple mathematical model of ski jumping should (if it has any merit) be able to reproduce them.

What about jumper weight? We found that for bobsleds, increased weight led to faster runs, and so bobsledders added weights up to the legal limit. For ski jumpers, it seems, reduced weight is preferable—it reduces

10. For theoretical calculations, computer simulations, and wind tunnel tests of ski jumping, see Luhtanen (1996); Maryniak, Ladyzynska-Kozdras, and Tomczak (2009); Müller (2009); Remizov (1984); and Ward-Smith and Clements (1983).

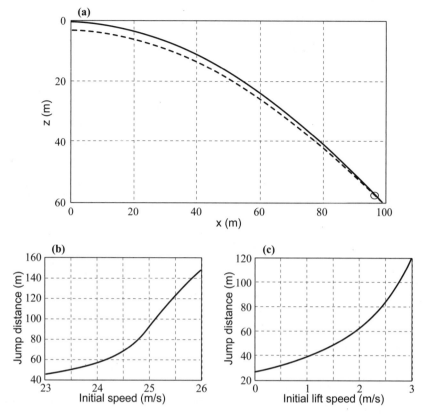

Figure 6.14. Simple analysis of ski-jump trajectory. (a) Trajectory shape (solid line) assuming an initial speed of 25 ms^{-1} and an initial lift speed of 3 ms^{-1}. The hill shape is shown by the dashed line. Note the very small intersection angle between the two shapes, ensuring a soft landing (unless the jumper falls). The K-point is shown by the open circle. (b) Jump distance vs. initial speed. (c) Jump distance vs. initial lift speed.

the relative importance of the downward force of gravity without reducing lift, and so flight distances increase as weight decreases. In the past, this fact has resulted in problems that affected ski jumpers' health.[11] Can a

11. Several cases of eating disorders (anorexia nervosa and bulimia) have been reported among ski jumpers who wanted to shed that last ounce to help get them onto the podium. To combat this sad trend, the FIS changed the rules governing ski length. The length of a jumper's skis (which influences aerodynamic lift) was previously limited by a formula that depended upon the jumper's body mass index (mass or weight divided by the square of his

simple mathematical model reproduce this effect—that reduced weight means increased distance, all else being equal?

Technical note 17 provides a simplified analysis of ski jumping, based on the fiction that the jumper is a simple point mass (like a cannon ball) who flies through the air with a constant lift-to-drag ratio. We have seen that jumpers are in fact able to adjust their configuration midair to render their lift-to-drag coefficient approximately constant, to their benefit and despite the ever-changing angle of attack throughout the flight. So this assumption (which is a huge simplification) is not too unrealistic.

In technical note 17 I also show how the hill shape can be specified in terms of a ski jumper's trajectory. I give it the same shape as a typical trajectory, but corresponding to an initial takeoff speed that is a little faster than most jumpers will attain. This ensures that the hill is almost the same shape as the trajectory of a jumper, so that he does not have too far to fall. To ensure that he does land, we say that the slope of the hill beyond the K-point is constant (and even farther away it levels off, of course). The result of this analysis is plotted in figure 6.14 for typical jumper speeds and aerodynamic parameter values.

Particularly noteworthy is the fact that when the jumper is airborne, he is never very far above the hill: the hill has been designed to ensure that this is the case. Calculations also show how the jumper's speed varies through the air. He slows down shortly after takeoff because of the drag force, but then gravity kicks in and pulls him downward with ever-increasing speed.

So far so good, but does this simple analysis reproduce the observed phenomena of ski jumping discussed earlier? Mostly. Unsurprisingly, we find that a jumper jumps farther if he takes off faster. This result seems to contradict the large-hill data from Vancouver 2010 that we plotted in figure 6.10a, which showed no discernible trend in the jump distance as speed varied, though the range of jump speeds was quite small (2 kph, or about 0.5 ms^{-1}). The graphs are different in another way: in figure 6.14b there is a clear trend, a smooth line: faster speed means longer distance. In figure 6.10a the data was all over the place. Why? Because our simple calculation assumes that all the jumpers have exactly the same lift and drag characteristics and exactly the same techniques. The real world is messier

---

height). In practice this formula benefited very lightweight jumpers—hence, the problems. A change to the rules has reduced the advantage for very underweight skiers. See Müller (2009) and Schmölzer and Müller (2002).

EDDIE THE EAGLE AND TOMMY THE TURKEY

Eddie "The Eagle" Edwards competed for Great Britain in the 1988 Winter Olympics in Calgary. He was a capable downhill skier who did not quite make the cut for the British Olympic team in 1984, so for Calgary he was determined to try his luck in ski jumping, a discipline in which he had no previous experience. There were no ski-jumping hills in Britain and no ski jumpers, apart from Eddie. He therefore had little difficulty in making the cut—but a lot more difficulty in making the jumps. He was too heavy and wore glasses that fogged up; he was self-financed and so made do with ill-fitting, second-hand equipment and indifferent technique. Nevertheless, he achieved his goal of setting a British record, of 73.5 m, despite finishing last.

His very amateurishness, coupled with an engaging personality, endeared him to the general public but not to the sport's governing body. The FIS changed the rules, incorporating the "Eddie the Eagle Rule," which barred entry to competitors who were not among the top 50 jumpers in international competitions. You may regard the FIS as guardians of the sport or as spoilsports, depending on your answer to this question: Is entertainment more or less important than achievement?

Now consider Tommy the Turkey. He differs from Eddie the Eagle in two important ways. First difference: Tommy is a figment of my imagination. He is an American skier who has qualified for the U.S. ski-jump team by virtue of his splendid jumping technique. This pristine technique, and his top-of-the-range equipment, is the second difference. He is also a gastronome who, shortly after qualifying for Olympic competition, developed a passion for deep-fried chocolate bars. The inevitable consequence, which caused shock to his trainer and teammates when he showed up at the Olympic venue, was that Tommy had ballooned out to a massive 150 kg weight (that's 330 lb), which was exactly twice his qualifying weight. The jump that we plotted in figure 6.14a was Tommy's qualifying jump, but in the Olympic competition, he could get nowhere near this distance because of his increased weight. Everything else was the same: he achieved the same take-off speed and initial lift speed, and he was able, through his splendid flight technique, to maintain the same lift-to-drag ratio as he did during qualification. The only difference in his jump parameters was his weight. So how far did Tommy jump? Forty-three meters. He then fell over and rolled down the hill, generating a massive snowball that knocked down all five of the judges.

This is why ski jumpers tend to be skinny.

than that, and variation in jumper characteristics (apparent in fig. 6.10a) masks completely the predicted variation of jump distance with takeoff speed. In figure 6.14c we also see that jump distance increases with initial lift speed, in accordance with observation of real ski jumps.

What about jumper weight? Here the model is still in agreement with observation: reduced jumper weight results in increased jump distance, all else being equal. We can see why, by considering the flight of eagles and turkeys.[12]

12. There is one feature of the simple analysis of technical note 17 that is not quite right. For real hills, the K-point occurs at an angle of about 37°. That is, the slope of the hill is about 37° at the "par" landing position. Data from the FIS Web site is remarkably consistent on this point. (For the 10 jumping hills that are specified, the slope at the K-point varies from 34.9° to 38.0°.) For the model of technical note 17 the corresponding angle is a little larger, at 45°. Probably this is because the FIS specifies hill shapes in a different way than I do in technical note 17.

THE FINISH LINES

For an eloquent graphical statement about the eagerness and enthusiasm with which we embrace winter sports, and about the drive which propels athletes to greater heights (actually, lesser heights in downhill events), see the graph on p. 142. This graph shows how the world record time for the men's 1,500-m long-track speed skating event has decreased over the years. In something more than a century, the time taken for the world's fastest skater to cover a distance that is only a little short of a mile has been reduced by a third. Several of the earliest records were set by Peder Østlund of Norway; the latest (as I write these words) by Shami Davis of the United States. These skaters, and all those between them from many different countries who contributed to this graph, share at least one thing: a strange and all-consuming desire to be the fastest man on ice.

The desire to contribute to a point on this graph is strange because, well, why should anyone feel the need to strap metal blades to their feet and career rapidly over a surface of solid water? It is all-consuming because, these days, athletes who aspire to world records must devote a significant part of their youthful years to their goal: most of their time for several years is taken up with constant training to improve strength, endurance, and technique. Consider the human effort that must have been exerted to contribute one point of the graph. Each point represents not just the exertions of the skater during the race in which he set the world record, but also the exertions of all the other athletes in that race, who were pushing him on, forcing him to his best efforts, goading him forward with the last ounce of strength to crash over the line in record time. Similar exertions were made for all the earlier races that led up to this moment. Now look at how many points there are on the graph, contributed over

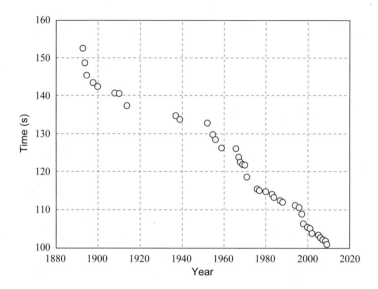

Men's 1,500-m long-track speed skating record time, 1880–2010.

many years, and you can see that hundreds or thousands of people have spent, in total, centuries of extreme effort to make this graph.

Every other distance in long-track speed skating can produce a graph like this; every other winter sport has its own set of graphs for each type of event. I have already noted that in the Vancouver Winter Olympic Games, men's cross-country skiing (the 30-km pursuit won by Sweden's Marcus Hellner) all the medal winners and many of those who finished out of medal contention literally fell to the ground in utter exhaustion immediately after crossing the line. In the women's cross-country sprint event at the same games, the eventual bronze medal winner recovered from a bad fall during warm-up before the race to ski the entire race with four broken ribs and a collapsed lung. (Petra Majdic of Slovenia skied against medical advice and in great pain; she did not know the extent of her injuries at the time.)

It stirs the blood of even an armchair winter sports enthusiast such as me to see people rise to the challenge in these competitions. At the time of writing, the Vancouver Winter Olympics are still fresh in the memory: who can forget the courageous Joannie Rochette, who won a bronze medal in figure skating just days after the death of her mother? Or Apolo Ohno winning an eighth short-track speed skating medal—a record number? Or the wave of national euphoria and relief as the Canadian men narrowly defeated the surprisingly good (to all but themselves) Americans in the men's hockey final?

Anybody who watches winter sports can engage with the emotions that accompany the triumphs and tragedies of individual athletes. However, as a physicist I find that my appreciation of the achievements of athletes in setting world records or in overcoming adversity is enhanced by an understanding of the basic physics that underpin the different sports. I can appreciate the drama at an emotional level as much as other people do, but my knowledge of physics permits me to see the achievements in a different light, or from another angle. I hope that, having read this book, you too will have gained an enhanced appreciation of winter sports.

Consider the graph again. Note that the reduction in race time is not uniform: there are two near-plateaus, and three regions of rapid reductions when the world record tumbled frequently. Why such uneven progress? The sharp drop in the late 1990s we can attribute to the widespread use of clap skates. What about the drop in the late '60s? Is that due to emerging talent or technology? What about the initial rapid drop in race times at the end of the nineteenth century? Is this to be expected, simply on statistical grounds, at the beginning of a recorded sequence, or is there something else going on? I will leave you to ponder these questions, and others that may have arisen in your mind from the analyses presented in this book. Looking at winter sports with a physicist's eye raises deeper questions as well as supplying answers.

PONDERABLES

Here, I provide some questions for you to ponder. Regard these as tutorial questions if you are a university student, or as brain fodder if you are reading this book simply to scratch an intellectual itch. Some questions are easy, others are hard. Some are short, others long. In some cases I will provide a hint or partial answer to the question posed; in other cases you are completely on your own.

• *The hockey stop.* Check out the YouTube video at www.youtube.com/watch? v=br8dfnnWL5k&feature=related, which shows how hockey players can stop quickly. They lean backwards and slide with their skate blades perpendicular to their direction of motion. Recall from chapter 2 that moving an ice skate in this direction is very difficult; blades are designed so that they won't move perpendicular to their length. Note the large amount of ice chips sprayed out in front of the rapidly decelerating player. (From the video, it is clear why the ice is resurfaced between periods of a hockey game.) Some professional hockey players make use of this spray to put the opponent's goalie at a disadvantage during a game. Assume that movement over the ice during a hockey stop can be modeled as sliding friction with a large coefficient of friction. From the video, estimate the angle at which the hockey player is leaning back (to prevent himself going head over heals) and from this angle, determine the friction coefficient. If the player's initial speed was 10 ms^{-1}, how long will it take him to stop?

• *Skate design.* Check out the design of a speed-skater's boot online. Because speed skating events (on both long and short tracks) proceed in the counterclockwise direction, skaters going around the bends must pass the right skate over the left, but never the left skate over the right. How does this influence the boot design? Contrast speed skaters' footwear with ice hockey or figure skating skates: in both these sports the left skate may cross over the right skate as much as the other way round.

• *Drafting.* We looked at speed-skating statistics from Vancouver's 2010 Winter Olympic Games in the figure on page 42. These statistics (medal winners' times and the distances covered in each long-track event) are readily available online. How did I estimate athletes' relative power output from such data? The energy expended by each athlete in skating a distance x is approximately $E = \frac{1}{2}mv^2 + mbv^2x$. The first term on the right side is the skater's kinetic energy; the second term is the work he or she expends in overcoming aerodynamic drag. The drag factor for a single skater is about $b = 0.001$ m^{-1}; you will find that it is ⅙ less for a team pursuit skater over the course of a race. Show from this (assuming that each skater does his fair share of time leading the team) that, when drafting, a team skater expends only ¾ of the average power of a skater who is on his own.

• *Curling.* Derive a natural (realistic and plausible) theory of curling. Award yourself a master's degree in physics.

• *An altitude problem.* The analysis of technical note 10 led to an equation that describes how race times vary with altitude. There are a number of simplifications that I made for this analysis but did not state explicitly. (For example, I tacitly assumed that an athlete's speed doesn't vary much during a race, so that it makes sense to express speed as distance divided by race duration.) Describe some of the other assumptions. Given the simplifications, why is the result believable? Much of applied physics involves making simplifying assumptions that get to the core of a problem without cutting out the essentials—of knowing when and how to simplify. Derive the result, equation (N10.4), from the earlier equations.

• *Biathlon blues.* As we saw in chapter 1, the Winter Olympics sport of biathlon is a combination of cross-country skiing and rifle shooting. The athletes swarm around a course on skis, and then remove their skis and run to a rifle range, where they shoot at a target, before repeating the process. If their shooting score is low, they must do extra laps. The winner is the first skier across the finish line. Given the exhaustion that accompanies cross-country skiing (see chapter 5), we can expect that a competitor who is about to shoot will be out of breath—adversely affecting his aim. What is his best strategy? Does he wait to regain his breath, thus improving aim and so reducing the possibility of penalty laps, or does he take his chances and shoot as soon as he can, thus saving time (perhaps)?

• *Too much energy.* In note 15 to chapter 5, I point out that if we could turn off friction and drag, a skier heading down the slopes at Whistler Creekside would be moving at close to 300 mph at the finish line. Can you reproduce this calculation? Getting rid of dissipative forces simplifies the physics a lot. For instance, the final speed depends only on the height difference between start and finish (870 m in the case of Whistler), and not on the track length or shape.

• *Hitting the hill.* I mentioned in chapter 6 that the shape chosen for ski-jumping hills is based on reducing the speed perpendicular to the hill, so that the jumper will have a soft landing. In the simple math model of ski jumping in technical note 17, I chose a hill shape that was the same as the jumper's trajectory shape, but for a higher take-off speed. What other ways are there to specify the shape of the hill? The important point is to ensure that the hill is never very far below the jumper at any point of his trajectory. What about the shape of this hill beyond the K-point: how does this impose a maximum safe jumping speed?

TECHNICAL NOTES

These technical notes provide derivations for some of the equations that are graphed in the main text and back up certain claims that are made in the text. The level of math is variable—some of it is high school physics and math, and some is college level—and it is always condensed, to save space. However, if you want to reconstruct the derivations, there is enough development presented here so that you can fill in the gaps.

NOTE 1. FRICTION AND DRAG

Here I present the equations that we need elsewhere in this book for winter sports physics calculations. Sliding friction involves two solid objects in contact, moving with respect to one another. A typical classroom example is that of a block sliding down an incline, as shown in figure N1. If the incline angle, a, is carefully chosen, the block will not accelerate down the slope nor will it stick; it will slide down at a constant speed. In such a case there can be no net force acting upon the block, and so the forces shown in figure N1 must exactly balance. These forces are the force due to gravity (acting straight down), the force due to sliding friction (acting in the direction opposite to the block velocity), and the force N of the incline acting upon the block. This "normal" force points upward at an angle that is perpendicular to the slope. (The incline must exert a force on the block because the block does not fall through the incline, despite gravity.)

The equation describing sliding friction is

$$F_s = \mu N, \tag{N1.1}$$

where μ is the coefficient of kinetic friction and N is the normal force (which, for the case of fig. N1, is given by $N = mg \cos a$, where m is block mass and g is the acceleration due to gravity). The inclined plane can be used to determine the

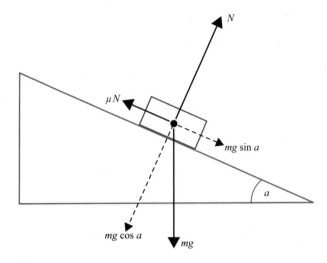

Figure N1. Sliding friction. A block sliding down an incline is subject to three forces (solid arrows). The gravitational force resolves into two components (dashed arrows) that are parallel and perpendicular to the incline; these components must equal in magnitude the normal force and the sliding friction force.

kinetic friction coefficient: because there is no net force acting on the block, you can see (by balancing the forces of fig. N1) that $\mu = \tan a$.

For lubricated friction, describing the physical situation that is illustrated in figure 1.6b, the friction force is expressed as

$$F_L = \frac{\mu_v A v}{h},\tag{N1.2}$$

where A is the slider area, v is slider speed, h is the thickness (assumed small) of the lubricating layer of liquid, and μv is the coefficient of dynamic viscosity of the liquid. Note that lubricated friction depends upon different parameters than sliding friction; in particular, it depends upon slider speed.

Aerodynamic drag is the friction force exerted upon a moving object (such as the puck of fig. 1.6c) by the air through which it travels. The magnitude of this drag force is

$$F_D = \tfrac{1}{2} c_D \rho A v^2.\tag{N1.3}$$

Here, A is projectile area (this time it is the cross-sectional area, as seen from the front), ρ is air density, v is projectile speed, and c_D is drag coefficient. Note that drag force depends upon the square of speed.

To describe all the forms of friction that act upon a winter sports athlete who is rapidly traversing a section of snow or ice, we will need either equation (N1.1) or equation (N1.2), plus equation (N1.3). Sometimes, if the athlete's speed is very low, we will not need (N1.3) because drag force will be small enough to be negligible. Sometimes we can ignore lubricated friction, equation (N1.2), because it is small in comparison to sliding friction. For many of our calculations we will have to take into consideration both sliding friction and aerodynamic drag.

NOTE 2. THE PUSH-OFF

To calculate the efficiency of a skater who pushes off from a standing start, consider the geometry of his skating motion, shown in figure 2.4b. I will make the simplifying assumption that friction is negligibly small for a skate moving along its length over ice. This assumption is reasonable for short steps such as occur when accelerating, though for longer, gliding strides I will need to include friction. The efficiency of our skater's motion is defined as

$$\epsilon = \left| \frac{P_{out}}{P_{in}} \right|. \tag{N2.1}$$

Here, P_{in} is the power that the skater expends: this is F^*v, where F^* is the force that the skater exerts against the ice and v is the skate speed. The output power, P_{out}, is the power used to drive the skater along the intended direction. To calculate this power, note that the force exerted by the skater along the direction of skate motion, indicated in figure 2.4b, is given by $F \sin 2a$. (This follows from the geometry, which I leave "as an exercise for the interested student," as physics professors like to say when they want their students to do some math.) The component of this force along the intended direction of motion is $F \sin 2a \cos a$, and so the output power is simply this force multiplied by skate speed, v. Thus, from equation (N2.1) we find the efficiency of our skater to be

$$\epsilon = \sin 2a \cos a. \tag{N2.2}$$

This efficiency attains a maximum value of 77%, for $a \approx 35°$.[1]

We can apply the same ideas to the backward-push motion shown in figure 2.4c. That noble creature "the interested student" can show from the geometry that a component $F \sin a \sin 2a$ of the reaction force is directed along the skate

---

1. We have calculated the *mechanical* efficiency. Of course, there are physiological factors that may also influence overall efficiency, such as muscle efficiency, arm length and movement, and so on. Our muscles are only about 20%–25% efficient, and so the power consumed by our muscles is four or five times as much as the mechanical power exerted.

motion, and a component $F \cos a \cos 2a$ is directed oppositely—causing the skate to slip. The net propulsive force is the difference between these two components, which is $- F \cos 3a$. A negative sign indicates negative acceleration—the skate slips. This occurs when angle a is less than 30°. The component of this net propulsive force that acts along the intended direction of motion is $- F \cos 3a \cos a$. Thus, the efficiency is $\epsilon = \cos 3a \cos a$, which peaks at 56% for $a \approx 52°$.

NOTE 3. BOOST-COAST SKATING

In figure 2.6 we saw how a skater's speed changes with time, depending upon his actions: if he is accelerating at the start of a race, then his stride rate is much greater than if he is gliding along at a constant average speed. Which is the best strategy for a skater who is trying to maintain speed, after the initial burst of energy? Should he skate with many short strides, or should he skate with fewer but longer strides? The answer depends upon the type of race.

To maintain speed, a skater boosts his speed with a power stroke (the upward sloping sections of the curves in fig. 2.6c) and then coasts until friction and drag reduce his speed to some minimum acceptable value, at which point he boosts speed again with another power stroke. During the boost phase, the force exerted by the skater against the ice is related to his speed over the ice via

$$F \tau \approx m(u - v_0). \tag{N3.1}$$

Here F is the force, m is skater mass, τ is the duration of the boost phase (i.e., of the power stroke), and v_0 and u are the skater's speeds at the beginning and end of the power stroke. The notation is made clear in figure 2.6c. Equation (N3.1) is only approximately true, but this approximation is good enough for our purposes.

During the coast phase of the skating stride, we must include the influence of sliding friction and aerodynamic drag. The equation that describes skater speed during this phase is

$$m\dot{v} = - \mu mg - mbv^2, \quad \text{where } b = \frac{c_d \rho A}{2m}. \tag{N3.2}$$

Here the first term on the right is the sliding friction force, and the second term is the aerodynamic drag, written as $F_d = mbv^2$ [compare eq. (N1.3)]. The constant b, here called the *drag factor*, has units of inverse length. In equation (N3.2) I am employing Newton's dot notation to indicate time derivative, so that $\dot{v}$ means rate of change of speed, v, with time, which is acceleration. Equation (N3.2) is easily solved for the skater speed during this coasting phase of the stride:

$$v(t) = \frac{u - q \tan bqt}{q + u \tan bqt} \, q, \qquad \text{where } \dot{q}^2 \equiv \frac{\mu g}{b} \,. \tag{N3.3}$$

I will assume that this coast phase lasts for a time t_0. At the beginning of the coast, you can see from equation (N3.3) that $v(0) = u$, which is correct (see fig. 2.6c). At the end of the coast we require the speed to be v_0, and so

$$v(t_0) = v_0. \tag{N3.4}$$

Now let's consider the energy expended by the skater during the boost phase. This energy is given by $E = F\bar{v}\tau$, where $\bar{v}$ is the average speed during the boost phase. The distance traveled during the stride—the stride length—is $\bar{v}(\tau + t_0)$, and so the energy expended per unit distance traveled is given by

$$\bar{F} = F \frac{\tau}{\tau + t_0} \,. \tag{N3.5}$$

We see that energy per unit distance is just the average force exerted by the skater over the duration of a stride. This is why I denote it $\bar{F}$. Substituting into equation (N3.5) from equations (N3.1)–(N3.4), we can plot $\bar{F}$ as a function of t_0. This plot is shown in figure 2.8a for realistic parameter values. It shows that for long races, in which the skater wants to minimize energy expenditure for each stride, he should take long strides—and this is what skaters do.

Now let us calculate the energy expended per unit speed, instead of per unit distance. We find

$$\frac{E}{\bar{v}} = m(u - v_0). \tag{N3.6}$$

Equation (N3.6) is plotted in figure 2.8b, which shows that, for short races where the skater wants to maintain high speed for minimum energy expenditure, he should take short strides—and this indeed is what happens. So, our simple physical model of a skater's stride is capable of explaining the difference in average stride lengths that occur in different types of races.

NOTE 4. SKATING A BEND

A skater hurtles round a tight bend at high speed—he must lean into the bend to avoid falling over—to resist the centrifugal force that pushes him away from the center of the bend. From figure N4 you can see that the total force that acts upon our skater, the vector sum of gravitational and centrifugal forces, will be directed down through the skate if the following condition is satisfied:

$$\tan \theta = \frac{v^2}{gR}, \tag{N4.1}$$

where R is the bend radius. If this condition is not satisfied—if the total force acts inside or outside the skate—then our skater will either fall into or fall out of the bend. For example, if we choose a realistic value for speed, say $v = 10$ ms^{-1}, then our skater will lean over about 50° when going around an 8-m-radius short-track bend. This lean angle is realistic. If our skater takes a 25-m long-track bend at the same speed, he will need to lean only about 20°.

Figure 2.9 shows the geometry of steps—the arrangement of straight-line strokes over the ice—that a skater might make in order to skate around a bend. From this geometry we can see that the step length l is given by, approximately,

$$l \approx \sqrt{2Rd}, \tag{N4.2}$$

where d is the sidestep length. We can estimate the sidestep length by counting the number of steps that a skater makes when skating in this manner around the 180° bends at each end of a speed skating track. Let us say that a skater takes N steps to completely go around the bend (so to speak). From figure 2.9 we see that $N \approx \pi R/l$ and so, from equation (N4.2) we see that

$$d \approx \frac{\pi^2 R}{2N^2}. \tag{N4.3}$$

Checking video footage of speed skating events, I count about 10 steps for a skater to go around an 8-m-radius short-track bend, which gives us $d \approx 0.4$ m. I count 16 steps for skaters to go around a 25-m-radius long-track bend, which yields $d \approx 0.5$ m. These numbers are similar, and look to be about right. Substituting back into equation (N4.2) shows that the stroke length around a short-track bend is about 2.5 m, and about twice as long for a long-track bend.

What force must a skater exert to power his way around a bend, maintaining constant speed? The forward component of force must equal the sum of sliding friction and aerodynamic drag forces, if speed is to be maintained, and so

$$F \sin a = \mu mg + mbv^2. \tag{N4.4}$$

The right side of equation (N4.4) is the force that a skater must exert to maintain speed on the straight[2] [see eq. (N3.2)], and so the force exerted around a bend is

---

2. This is not quite true because the sliding friction coefficient around a bend is about 35%–40% larger than the sliding friction on the straight, due to skating action. See Jobse et al. (1990).

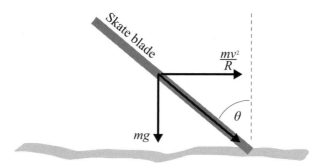

Figure N4. A skater leans over at an angle when taking a bend, so that the total force (gravity plus centrifugal) acts down through the skate (thick gray line).

greater than that on the straight. Given that $\sin a \approx \pi/N$ radians, we see that the skater's force around a short-track bend ($N = 10$) is about three times the force needed to maintain speed on a straight section of track. Around a long-track bend ($N = 16$) it appears that the required force is even greater; however, on such gentle bends it is likely that the skater can exert a force that is not directed toward the circle center, which I have assumed in figure 2.9, but instead includes a backward component, as happens on the straight. Only on tight bends do I expect the force to be significantly elevated.

Recall from chapter 1 and equation (N1.1) that sliding friction force increases with the normal force. Figure N4 shows how far a skater must lean over to counter centrifugal force, resulting in an increased force exerted by his skate upon the ice. From figure N4 the magnitude of this normal force is[3]

$$N = \sqrt{(mg)^2 + \left(\frac{mv^2}{R}\right)^2}. \tag{N4.5}$$

Thus, the friction force increases around tight bends and at high speeds, even if the skater is not powering his way around the bend but is simply gliding on one foot.

NOTE 5. IN A SPIN

We can gain insight into the spin elements of figure skating with the help of elementary physics. First I will calculate the height to which a skater must jump in order to complete four rotations before landing. A skater jumping to height h_n

3. Here, I am assuming that the skate digs into the ice, so that the normal force is directed at an angle θ to the ice surface (see fig. N4) and is not perpendicular to the ice surface, as the word "normal" implies. This is a reasonable assumption.

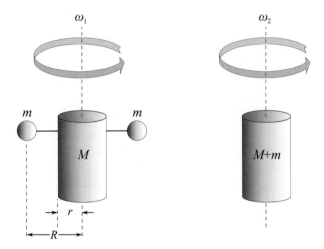

Figure N5. A skater spinning with arms extended (*left*), here modeled as a mass, M, with two attached masses, m, all spinning about the central vertical axis; and with arms pulled in (*right*).

will take a time t_n to do so, where h_n and t_n are related via $h_n = \frac{1}{2}gt_n^2$. He will then take the same time to fall back onto the ice. During the interval $2t_n$ when he is airborne, the skater executes n rotations at frequency f, so that $t_n = n/2f$. Thus,

$$h_n = \frac{gn^2}{8f^2}\,.$$

(N5.1)

For $n = 4$ we have a quad jump; for $n = 3$ we have a triple. Equation (N5.1) is plotted in the figure on page 49.

We saw in chapter 2 that a skater can use angular momentum conservation (angular momentum is almost a constant, if frictional forces are small) in order to increase angular speed during a spin. This is easy to demonstrate with a simple model. Figure N5 shows an unflattering (but reasonable, to a physicist) model of a skater with her arms extended and with her arms down by her side. We say that she has an angular speed of ω_1 when her arms are extended, and we want to determine her angular speed ω_2 when her arms are drawn in.

Angular momentum, J, is the product of moment of inertia and angular speed: $J = I\omega$. If moment of inertia and angular momentum do not turn your handle (another near pun), you may skip this calculation. Let me assume that angular momentum is approximately conserved during an ice skater's spin, so that

$$I_1\omega_1 \approx I_2\omega_2.$$

(N5.2)

The moment of inertia for the skater with her arms extended (on the left of fig. N5) is

$$I_1 \approx \tfrac{1}{2}Mr^2 + 2mR^2, \tag{N5.3}$$

whereas her moment of inertia with arms by her side is

$$I_2 \approx \tfrac{1}{2}(M + 2m)r^2. \tag{N5.4}$$

Let us choose nominal values of $M = 45$ kg and $m = 2\tfrac{1}{2}$ kg, $r = 0.2$ m and $R = 0.5$ m. Then from equations (N5.2)–(N5.4) we see that $\omega_2 \approx 2.15\,\omega_1$. So our model says that a skater can more than double her angular speed during a spin by drawing in her arms. I might extend the model to include a drawn-in leg, but I think you get the picture already: a leg would increase the amplification of spin speed.

NOTE 6. SLIPPERY SLOPE PHYSICS

We have seen that a skier or a bobsled sliding down a slope is subjected to three forces: gravity, sliding friction between snow and ski or between ice and sled rails, and aerodynamic drag. Gravity acts vertically downwards; contact friction and drag act in the direction that opposes movement—in other words, in the opposite direction to velocity. Here, we derive the formula that describes how a mass (a skier or a bobsled) moves when subjected to these forces, assuming that the gradient is constant (so that the slope angle—let us call it a—does not change with position on the slope). Perhaps surprisingly, this problem can be solved exactly (*analytically*, the mathematicians would say, or *in closed form*).

I set up the problem in figure N6. You can see that contact friction and aerodynamic drag act in the same direction, but from chapter 1 and technical note 1 we know that they assume different forms. The force of sliding friction takes the following form (see technical note 1):

$$F_s = \mu N = \mu mg \cos a. \tag{N6.1}$$

Aerodynamic drag is different because this force depends upon how fast the mass is moving:

$$F_d = mbv^2. \tag{N6.2}$$

In equation (N6.2) v is the speed of the moving mass and b (recall from technical note 3), is a constant drag factor. It is proportional to the aerodynamic drag

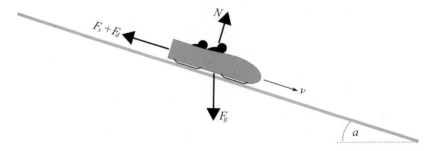

Figure N6. A bobsled moves with speed v down a straight track of constant gradient a, acted upon by the four forces shown: gravity, F_g; the normal force, N; sliding friction, F_s; and aerodynamic drag, F_d.

coefficient c_d; to the density of air, ρ; and to the cross-sectional area of the mass (i.e., how big the skier or bobsled looks from the front). Drag coefficient is here treated as constant; this is a pretty good approximation for most winter sports.

From figure N6 and from equations (N6.1) and (N6.2) we can write the equation of motion that describes the slippery slope physics:

$$m\ddot{x} = mg \sin a - \mu mg \cos a - mb\dot{x}^2. \tag{N6.3}$$

I am again using Newton's dot notation to indicate time derivative, so that $\ddot{x}$ is acceleration and $\dot{x}$ is velocity (actually speed, which is the magnitude of velocity, here understood to be in the downhill direction), so that $v = \dot{x}$. Equation (N6.3) can be solved for speed:[4]

$$\dot{x} = \sqrt{\frac{g}{b} (\sin a - \mu \cos a)(1 - \exp(-2bx)) + v_0^2 \exp(-2bx)}. \tag{N6.4}$$

Here, v_0 is the initial speed, corresponding to $x = 0$. This equation applies if the slope is steep enough to overcome static friction, so that the mass moves downhill instead of standing still. Mathematically, equation (N6.4) is valid for

$$\mu < \tan a. \tag{N6.5}$$

You can see from the form of equation (N6.4) that speed increases as the mass moves down the slope, approaching a terminal speed v_T given by

4. For the mathematically adept reader I note that the differential equation (N6.3) is separable and the resulting integration is elementary.

$$v_T = \sqrt{\frac{g}{b}} \, (\sin a - \mu \cos a).$$

(N6.6)

Why is there an upper speed limit? As the mass moves down the slope, speed increases (due to the force of gravity) until it is great enough that aerodynamic drag plus contact friction exactly matches gravity. [Recall, from eq. (N6.2), that aerodynamic drag increases with speed.] At this point the net downhill force acting on the skier or the bobsled is zero, so no further change in speed is possible until something else is altered, such as brakes being applied. The terminal speed is analogous to that of a skydiver, except that in our case there is an extra complication arising from the presence of contact friction.

From equation (N6.4) we can integrate to obtain position x as a function of time. For the sake of completeness, and to satisfy mathematically inclined readers, here is the solution:

$$x(t) = \frac{1}{b} \ln\!\left(\cosh(bv_T t) + \frac{v_0}{v_T} \sinh(bv_T t) \right).$$

(N6.7)

If hyperbolic trigonometric functions do not ring your chimes, then look instead at figure 3.5a in the main text, which shows a graph of the solution. Let us say that the skier or the bobsled travels a total distance of L down the slope and that it takes a time t_0 to do so. Thus, $L = x(t_0)$. Equation (N6.7) can be inverted to give us the descent or run time for the run of length L:

$$t_0 = \frac{1}{bv_T} \ln\!\left(\frac{v_T}{v_T + v_0} \exp(bL)\!\left(1 + \sqrt{1 - \left(1 - \frac{v_0^2}{v_T^2}\right)\exp(-2bL)} \, \right) \right).$$

(N6.8)

It is useful to obtain run time—a key statistic in many winter sports—in closed form like this because it allows us to see how the time changes with changing parameters. Thus, we can play around with the initial speed, v_0, or with the aerodynamic drag factor, b, to see what influence these parameters have. In figure 3.5 I have plotted the dependence of descent time (run time), t_0, on initial speed, drag factor, and mass or weight.

Of course, real luge or bobsled runs are not of constant slope and are not straight, but this analysis still provides a useful indicator of how important initial speed is. Anything that can be done to cut down aerodynamic drag (i.e., to reduce b) will help reduce run time.

To completely solve the problem I have set up, we must consider what happens when the inequality (N6.5) is not satisfied. In other words, we must solve the equation of motion (N6.3) for the case of a large friction coefficient or a small slope angle, such as the flat ice of ice skaters. In this case there is no terminal speed; the skater starts at speed v_0 and slows to zero. Her total distance traveled is[5]

$$L = \frac{1}{2b} \ln\left(1 + \frac{bv_0^2}{g(\mu \cos a - \sin a)}\right). \tag{N6.9}$$

We have seen that the results obtained here are applicable to different winter sports in different ways. So, for example, they apply to an ice skater if she is moving in a straight line and if we set the slope angle a to zero. They apply approximately to the along-ice component of the motion of curling rocks (though not to the across-ice component). The equations apply to luge, skeleton, and bobsled on sections of the descent that are of constant slope and straight, with initial speed v_0 and initial position $x = 0$ corresponding to the beginning of this straight section. With a different kinetic friction coefficient these equations apply to downhill skiers. We can apply the results obtained here to many of our winter sports, with suitable adaptations.

NOTE 7. BOBSLED AND LUGE STARTS

For two-man bobsled, measurements of push-off times and speeds have shown an empirical relation between the two.[6] In other words it has been observed that the time (τ) taken for a two-man bobsled to reach the start line is related to the speed of the sled at the start line, as follows:

$$v_0(\tau) = A - B\tau. \tag{N7.1}$$

Here, $A = 22.82$ ms^{-1} and $B = 2.06$ ms^{-2}. The power generated by bobsled athletes at the push-off, which accelerates themselves and their sled, is $P = W/\tau$, where W is the work done in moving the athletes and sled from their initial position to the start line. Thus,

$$P = \frac{1}{\tau} \left[\tfrac{1}{2}m(A - B\tau)^2 - mgl \sin a\right], \tag{N7.2}$$

where m is the mass of athletes plus sled, $l = 50$ m is the distance to the start line from the push-off, and a is the track gradient or slope. It is important to note that

5. In more elementary treatments aerodynamic drag is ignored, in which case the total distance moved is $L = v_0^2/2\mu g$. The mathematical reader can show that equation (N6.9) reduces to this simpler form in the limit $b \to 0$.
6. The data below is from Lewis (2006).

P is the total power; the power expended by each athlete is P divided by the number of athletes pushing the sled. Equation (N7.2) is plotted in figure 3.6a.

Substituting for τ in equation (N7.2) from (N7.1), and inverting to obtain start speed in terms of power, we find

$$v_0 = \sqrt{\left(\frac{P}{mB}\right)^2 + 2A\left(\frac{P}{mB}\right) + 2gl\sin a} - \left(\frac{P}{mB}\right). \qquad (N7.3)$$

Equation (N7.3) is plotted in figure 3.6b.

In writing down the equations that govern the luge push-off phase, we can ignore aerodynamic drag because the initial speed in this sport is so small that drag is unimportant during the first second or so. Let us say that the arms and upper body of a luge athlete can apply 500 W of power, P, during the pull which lasts, let us say, for one second ($\tau = 1$ s). The power exerted is the energy that it gains during this pull, divided by the time taken:

$$P = \frac{\frac{1}{2}mv_0^2}{\tau}. \qquad (N7.4)$$

Here, v_0 is the luge speed that results from the pull. Inverting equation (N7.4) we estimate this speed to be $v_0 \approx 3.1$ ms^{-1} (about 11.3 kph). You can see that initial speed increases with the square root of power: more power means more speed. Even a small change in the initial speed can produce significant changes in run time, as we saw in technical note 6 (or fig. 3.5c); thus, in luge as in bobsled and skeleton, the push-off phase is crucial to success. Indeed, if we apply equation (N6.8) to the luge, we note that a reduction of initial speed from 11.3 kph to 11.2 kph results in an increase in run time of 0.035 second for the straight track. This is important in a sport that times runs to thousandths of a second.

NOTE 8. THE CENTRIFUGAL CONTRIBUTION TO SLED TRAJECTORY

Before we consider how centrifugal force slows a sled, we need to see how it works. Consider figure N8. Three forces are shown. The magnitudes of the gravitational, centrifugal and normal forces are, respectively,

$$F_g = mg$$

$$F_c = \frac{mv^2}{R} \qquad (N8.1)$$

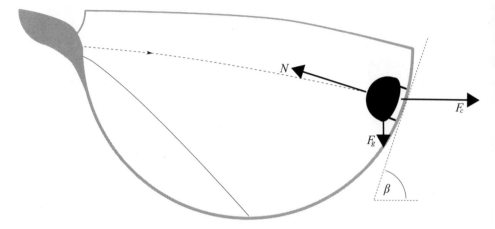

Figure N8. A sled rides up a bank to angle β, which depends upon sled speed and bend radius. In the text, β is calculated from the three forces shown: the gravitational force, F_g; the centrifugal force, F_c; and the normal force, N.

$$N = \sqrt{(mg)^2 + \left(\frac{mv^2}{R}\right)^2}$$

A few words of explanation are required for readers whose physics is rusty. (If your physics is uncorroded, or if you are not interested in the details, you may skip this paragraph.) The centrifugal force increases as the square of sled speed and is inversely proportional to the radius of the track (the curve radius, not the radius of the halfpipe). The sum of these two force vectors—gravity and centrifugal force—must be perpendicular to the track (the section of track beneath the sled). The reason is that if there were a component of the sum force that was parallel to the track, the sled would slip sideways under the action of this force. Runners may prevent sideways slipping, but the driver does not want much sideways force because it will increase friction and may cause the sled to overturn. So, we can reasonably assume that the sum of gravitational and centrifugal forces is perpendicular to the track. This force must be opposed by the normal force, which acts in the direction shown in figure N8. The sled pushes down on the track, and the track, being solid, pushes back. Thus, the normal force is a reaction force. The normal force must exactly balance the other two forces, because the sled does not sink into the track or rise above it. So, the normal force has the same magnitude as the vector sum of gravitational and centrifugal forces, as in equation (N8.1).

From figure N8 we can show that in order to be free of sideways force, the sled must be at an angle β (the "natural" angle for the sled when taking the bend), where

$$\tan \beta = \frac{v^2}{gR}. \tag{N8.2}$$

So, high speed or tight bends mean that the sled rides very high up the side of the bend.

When on a bend, the sled is subject to a sliding friction force of magnitude

$$F_s = \mu N = \mu mg \sqrt{1 + \left(\frac{v^2}{gR}\right)^2}. \tag{N8.3}$$

Tracks must be designed so that the normal force does not exceed five times the weight ($5mg$)—this is the 5-g limit mentioned in the text. So, the track must have a radius, R, such that the square root factor in equation (N8.3) is, at most, 5. The maximum speed for sleds is about 140 kph, which means that each bend must have a radius that exceeds 32 m (about 100 ft).

Now let us look at the friction force on a bend: it is increased by the g factor of the bend. Consequently, a sled is slowed down when it goes through a bend because friction is greater than for a straight section of track. Consider figure 3.9 of the main text. These two tracks are the same length and differ only in that one of them has a bend of radius R; the other track is straight. Let's estimate the loss of time down the curved track by considering the change of energy of a sled as it proceeds from the top to the bottom of each track. Assume that the two horizontal lines shown are contours separated by 100 m. The track with bends has two straight sections of length $R = 35$ m and a bend of the same radius; both tracks have a length of $s = 125$ m. Assume that the gradient is 5° so that the drop in height from the top to the bottom of each track is $h = 8.7$ m. Say the sled speed, when entering this section of track, is $v = 30$ ms^{-1}. Then the sled gains the following amount of energy when sliding down the straight track:

$$\Delta E_{straight} = mgh - \mu mgs - mbv^2s = +8{,}220 \text{ joules.} \tag{N8.4}$$

Here, I have used the values of μ and of b that we determined in chapter 3. This energy gain results in an increased sled speed, so that we can also write

$$\Delta E_{straight} = \tfrac{1}{2}m(v + \Delta v_s)^2 - \tfrac{1}{2}mv^2 \approx mv\Delta v_s. \tag{N8.5}$$

From equations (N8.4) and (N8.5) the increase in speed down the straight track is easily determined: $\Delta v_s = 0.435$ ms^{-1}. (I have assumed for the above analysis that Δv_s is much less than v, which we can now see is true.) Repeating this calculation for the curved track we find that $\Delta v_b = -0.018 \text{ ms}^{-1}$. In words, the sled loses a little speed as it heads down the curved track. This is because more energy is lost to friction on the bend than is lost on the straight sections of track; this energy loss is greater by a factor that is just the g factor, which we calculate from equation (N8.3). I leave this calculation of Δv_b once again to that semi-mythical person beloved of textbook authors, "the interested student." Also to this splendid person I leave the following simple calculation: converting the differences in speed to a difference in time traveling down the two tracks. This time difference is readily seen to be $\Delta t = 0.031$ s, as stated in the main text. Such a time difference corresponds to a distance of about 0.9 m (3 ft).

NOTE 9. CURLING ROCK CONCERT

Here I show how a simple model of sliding friction can account for the coupling between the rotational and the translational motion of a curling rock (that is, between its spinning and sliding). In figure 4.7 we saw the local velocity of an element of the rock running band as it slides over the ice: this velocity is the vector sum of center-of-mass velocity and rotational velocity at the running band element. The local friction force at the element of running band is of magnitude μmg and in the opposite direction to local velocity (here I assume that the friction coefficient is constant). You can see from figure 4.7 that the local velocity direction changes at different points on the running band.

Newton's second law leads us directly to the equation governing linear center-of-mass motion:

$$\dot{v} = -\mu g. \tag{N9.1}$$

Here, v is, as always, the speed (i.e., the magnitude of the velocity vector), so that $v = |\mathbf{v}|$. To determine the equivalent equation for rotational motion, I must once again invoke the "interested student" and invite him or her to confirm that the frictional torque that acts about the rock axis is

$$I\dot{w} = -\mu mgR \langle \mathbf{R} \times \hat{\mathbf{v}}(a) \rangle \text{ where } \hat{\mathbf{v}}(a) = (\mathbf{v} + \mathbf{w}(a))/|\mathbf{v} + \mathbf{w}(a)|. \tag{N9.2}$$

The angled brackets in this equation indicate averaging over the running band angle a of figure 4.7. If you are not in the mood for vector multiplication, we can skip to the equation governing rotational speed, derived from equation (N9.2):

$$\dot{w} = -\mu gn \frac{w}{v}, \qquad \text{where } n = \frac{mR^2}{2I}. \tag{N9.3}$$

In equations (N9.2) and (N9.3) I is the rock moment of inertia about its axis. For typical rock parameters, the constant n is about equal to 0.2. Recall that $w = \omega R$, where ω is rotation rate and R is running band radius, so that w is the rotational speed at the running band.

From equations (N9.1) and (N9.3) it is straightforward to show that

$$\omega = \omega_0 \left(\frac{v}{v_0} \right)^n. \tag{N9.4}$$

The subscript 0 indicates initial value. Equation (N9.4) shows us how linear and angular speeds are linked. Because n is less than 1, angular speed slows down much more gradually than linear speed, except at the very end of the trajectory. [The numbers illustrating this, given in chapter 4, come from equation (N9.4).] From this equation we can also see that linear and angular motion cease at the same time, as we found for puck motion.

NOTE 10. GETTING HIGH ON SPEED

I can provide a rough indication of the way that altitude influences an athlete's racing times with the following simple calculation.

Let's assume that an athlete taps energy E_{in} during a race, which is proportional to the oxygen that he or she consumes. We can then write $E_{in} = E_0 + k_0 \tau_0$. Here, E_0 is the energy (oxygen) that is stored inside the athlete at the race start, and $k_0 \tau_0$ is the energy (oxygen) taken in during the race, which is of duration τ_0. The constant k_0 depends upon the athlete's lung capacity and rate of breathing, as well as upon the altitude at which the race takes place. This energy is expended partly in accelerating the athlete but mostly in overcoming aerodynamic drag. We can write for the expended energy $E_{out} = \frac{1}{2}mv^2 + mbv^2 D$, where mbv^2 is the force required to overcome drag (see technical note 3) and D is the race distance. Equating input and output energies, and expressing athlete speed v as D/τ_0, we have

$$E_0 + k_0 \tau_0 = \frac{c_0}{\tau_0^2}, \qquad \text{where } c_0 = \tfrac{1}{2}mD^2 + mbD^3. \tag{N10.1}$$

At a higher altitude, the energy balance is different:

$$E_0 + k_1 \tau_1 = \frac{c_1}{\tau_1^2}. \tag{N10.2}$$

Let me say how I think the constants k_1, c_1, τ_1 are related to k_0, c_0, τ_0, and then explain why this is the case:

$$k_1 = (1 - \epsilon)k_0$$
$$c_1 = (1 - 0.8\epsilon)c_0 \tag{N10.3}$$
$$\tau_1 = (1 - \lambda\epsilon)\tau_0$$

Parameters $k_{0,1}$ describe the intake of oxygen during a race. The density of air falls by about 3% for every increase of 300 m (1,000 ft) in altitude, so the parameter ϵ is 0.03 if our higher-altitude race [described by eq. (N10.2)] takes place 300 m higher up than our lower-altitude race [eq. (N10.1)]. Parameters $c_{0,1}$ describe the energy required to overcome inertia and aerodynamic drag. We have already seen that about 80% of energy is needed to overcome drag; we can assume that the energy expended to overcome inertia is independent of altitude. Hence the second equation of (N10.3). We want to determine how race times $\tau_{0,1}$ are related, so we include a factor λ, to be determined.

From equations (N10.1)–(N10.3) we can show (after a little algebra, and assuming that $\epsilon \ll 1$) that

$$\lambda = \frac{0.8 - 0.2\alpha}{2 + 3\alpha}, \qquad \text{where } \alpha = \frac{k_0\tau_0}{E_0}. \tag{N10.4}$$

So our calculation shows that λ is positive if $\alpha < 4$. In other words, race times are reduced at higher altitude if the athlete's initial store of oxygen exceeds a quarter of the oxygen he breathes in during the race. In races where $\alpha > 4$, race times increase at higher altitude. Given the simplifying approximations that I have employed for this calculation, we should regard the result as only roughly correct. The general trend is plausible, however: in short races, where stored oxygen is an important factor, we can expect faster times at higher altitudes. In longer races, where oxygen breathed in during the race is more significant, the beneficial effect of high altitude is reduced or even reversed.

Not much data has been published on this subject, but we can glean some support for our conclusion by considering the 10 most recent world records set in long-track speed skating, at all distances, taking into account the altitude of the race venues. In figure N10, I graph the average altitude of the 10 world record venues against the race distance. Speed skating is the best sport to illustrate the influence of altitude because the race track shape and length is the same at all venues. (This is not the case for skiing events, for example.) This uniformity reduces the variability of results. Note the general trend: at short distances, world records are set at high altitudes, whereas for the longer races altitude is less significant. These data support the conclusion we drew from my simple calculation. To gain a more quantitative understanding would require a more detailed analysis and more detailed data, and I don't want to go there—this technical note is already long enough.

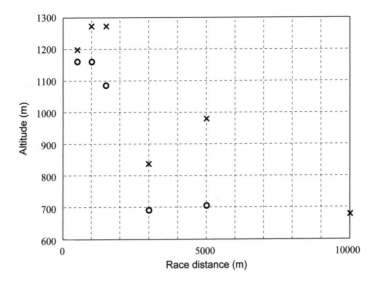

Figure N10. Average altitude of long-track speed skating world record events by race distance, for men's (x) and women's (o) events. For the 10 most recent world records, the venue altitudes have been averaged. For example, for the men's 500-m race, 6 of the last 10 world records were set at Calgary (alt. 1049 m) and 4 at Salt Lake City (alt. 1423 m). The average altitude is 1199 m. Data is from the International Skating Union Web site, *Historical World Records*.

NOTE 11. PARABOLIC SKIS

How do parabolic skis work? We can understand the basic idea from figure N11. First, the shape: parabolic skis are side-cut so that the edges, when the ski is flat, lie on a circle, as shown in figure N11a. From this figure and elementary geometry it is not hard to show that the side-cut radius C is given in terms of the other parameters by

$$C = \frac{L^2}{8D},$$
(N11.1)

where D is the side-cut distance and L is the length of the curved section of ski.

In fact, the shape is part of a circle only when the ski is flat; when it is cambered (i.e., when not loaded with the weight of a skier) the ski edge is not quite circular but is parabolic, as shown in figure N11b. When a plane intersects a cone, the line of intersection is a parabola, as you may recall from high school geometry. Step on the ski, and the parabola flattens into a circle, as indicated. The radius of this circle is clear from a consideration of figure N11c. Here we have a

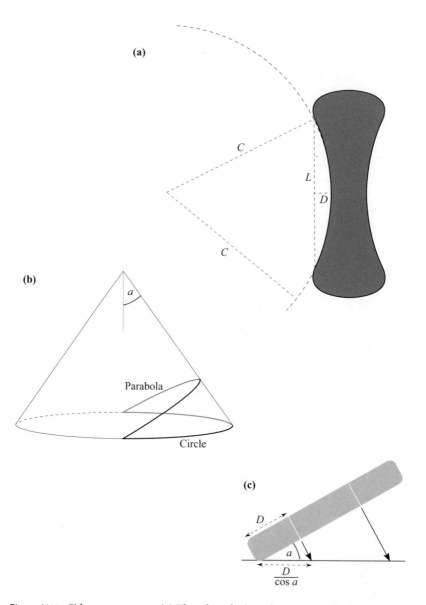

Figure N11. Side-cut geometry. (a) The edge of a (greatly exaggerated) side-cut ski is approximately on a circle. (b) More accurately, it lies on a parabola, defined as the intersection of a plane and a cone. When flattened by a skier's weight, the side-cut edge lies on a true circle. (c) The ski seen end-on; white lines indicate the ski edge near the binding. The horizontal line represents the depth to which the ski, inclined at angle a, is depressed (along its side-cut length, L) by the skier's weight. The radius of the true circle—of the curve carved by the skier—is smaller than the side-cut radius by a factor $\cos a$.

ski viewed end-on and inclined to the snow surface at an angle a. When the ski is pressed into the snow by the full weight of the skier (whose full weight is applied to one ski during a carve turn), the camber flattens out along the entire length of the ski. The geometry means that the side-cut depth is effectively increased from D to $D/\cos a$. Now look again at equation (N11.1): if D is changed during a turn, as indicated, then the side-cut radius changes to

$$R = C \cos a. \tag{N11.2}$$

Thus, the radius of the turn that is carved out by a skier depends upon the angle at which she leans.

We can also show that the radius of a curve depends upon the skier's speed. We saw earlier (e.g., fig. N4 or fig. 2.10) how a skater leaned over (inclined) when rounding a bend, in order to avoid being toppled by centrifugal force. The same physics applies here. If we neglect angulation, so that the ski angle a of figure N11c is the same as the lean angle θ of figure N4, the condition for a skier to avoid being tipped by centrifugal force is as follows:

$$\frac{mv^2}{R} \cos a = mg \sin a. \tag{N11.3}$$

This equation means that curve radius and lean angle are connected. We expect this from everyday experience: you know that you have to lean over more when cycling or skating around a tighter bend, or when taking a bend faster. But look what happens when we substitute equation (N11.2) into (N11.3) for the side-cut radius R:

$$v^2 = Cg \sin a. \tag{N11.4}$$

From equations (N11.2) and (N11.4) it is easy to see that curve radius and speed are related:

$$R = \sqrt{C^2 - \frac{v^4}{g^2}}. \tag{N11.5}$$

Equation (N11.5) shows us that a skier carves a curve much differently than a skater takes a bend. The skater takes a tight bend slowly and a shallow bend fast; equation (N11.5) tells us that the skier does the opposite. Tight curves are taken fast and shallow curves are crossed slowly. However, the skier's lean angle is greater for tighter bends, as for skaters, because

$$\sin a = \sqrt{1 - \frac{R^2}{C^2}} \, . \tag{N11.6}$$

Because of the different physics of carving, skiers had to learn a new technique, perhaps a counterintuitive one, when parabolic skis replaced the straight-sided classical skis. Note that there is a maximum radius of curve that can be carved out with a parabolic ski: it is C.

The analysis presented here assumes no angulation on the part of the skier—that is to say, she carves out a curve with a very straight body line. In practice, of course, she will lean over, tuck, or adjust the angle of her body line to the ski. In this way, she can modify the manner in which curves are carved from the behavior predicted in equations (N11.5) and (N11.6). For example, if she adjusts her orientation with respect to the ski by an angle ϕ, then equation (N11.2) is changed to $R = C \cos(a + \phi)$.

NOTE 12. ALPINE ENERGY EQUATION

We can write the energy equation of a downhill skiing event as follows:

$$mgh = \mu mgL + mbv^2 L + \tfrac{1}{2}mv^2. \tag{N12.1}$$

In words: the initial gravitational potential energy of the skier (with which he is imbued at the start of the race) equals the energy he spends in overcoming sliding friction as he skis down the course, plus the energy he spends overcoming aerodynamic drag on the course, plus the kinetic energy he possesses at the finish. I will apply this equation to the men's downhill final at the 2010 Winter Olympics in Whistler. In equation (N12.1) h is the height of the start line above the finish line (870 m in the case of Whistler), L is the length of the track (3,158 m), v is skier speed, and m is his mass. We will choose $m = 80$ kg (though this is not really important for our purposes, since the mass cancels) and $v = 27.5$ ms^{-1}, the average speed of the medal winners. To obtain the numbers quoted in chapter 5, I assumed a value of $b = 0.0014$, which is at the high end of the range of values measured by researchers (see the discussion in chapter 5). I chose this value because it corresponds to the heavy air conditions that applied to the Whistler Creekside course on the day of the downhill race (February 15; the race was delayed for two days because of rain and warm weather). For the same reason I chose a high value for the sliding friction coefficient ($\mu = 0.155$). Substituting these parameter values into equation (N12.1) yields the number quoted in chapter 5: most of the energy goes to overcoming friction and drag.

NOTE 13. CROSS-COUNTRY POLING

Poling is the method by which cross-country skiers scrabble over level ground, or up inclines: their awkward-looking running or sliding gait is assisted by ski poles. We can understand something about the physics of poling with a simple calculation. It involves a little math but gets us where we want to go. For the purpose of this calculation, we can assume that sliding friction dominates and so can ignore the effects of aerodynamic drag. (In chapter 5 we saw that sliding friction consumed at least 85% of the skier's energy budget.) Assume also that the ground over which our skier travels is level.

In figure N13 you see how speed changes throughout a stride cycle. For the force acting upon the skier, we write

$$
\begin{aligned}
m\dot{v} &= ma - \mu mg, && 0 \le t < t_0; \\
&= -\mu mg, && t_0 \le t < t_0 + t_1.
\end{aligned}
\tag{N13.1}
$$

Here, ma is the reaction force of the ground pushing against the skier (through her feet or her poles). The intervals $t_{0,1}$ are defined in figure N13. Integrating equation (N13.1) we obtain

$$
v_{n+1} = v_n + (a - \mu g)\, t_0 - \mu g t_1.
\tag{N13.2}
$$

In this equation v_n is the skier speed at the beginning of the nth stride. If our skier is progressing steadily over the level terrain, her speed will be the same at the start of each stride, so that $v_{n+1} = v_n = v$. Thus, from equation (N13.2)

$$
\Gamma = \frac{\mu g}{a}, \qquad \text{where } \Gamma = \frac{t_0}{t_0 + t_1}.
\tag{N13.3}
$$

The Greek capital gamma, Γ, is called the *duty factor* and is the fraction of time that the skier spends poling. The rest of the time she is gliding over the snow and repositioning the poles and her feet for the next stride. The *stride frequency*—the rate at which our skier strides (and the rate at which she poles over the surface)—is

$$
f = \frac{1}{t_0 + t_1} = \frac{\Gamma}{t_0}.
\tag{N13.4}
$$

The power expended by the skier each time she poles is the product of the force ma that she exerts during a stride, and the average speed v, so $P = mav$. Her power expenditure averaged over a stride cycle is

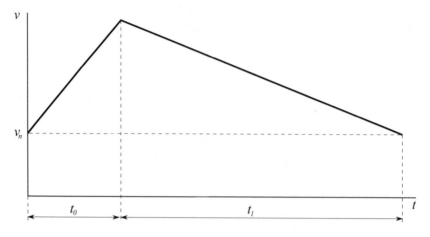

Figure N13. Speed vs. time in cross-country skier poling. Here, one stride is shown. Speed initially rises as the skier steps forward, assisted by poles. Then sliding friction causes the skier to slow down. We assume that the average stride speed is steady, so that it is the same at the end of the stride as it was at the start.

$$\bar{P} = \mu mg v = \Gamma ma v. \tag{N13.5}$$

Now we integrate equation (N13.2) again to obtain the *cycle length* (d), which is the distance moved by the skier while her ski poles are planted on the ground:

$$d = \tfrac{1}{2}(a - \mu g)t_0^2 + v t_0. \tag{N13.6}$$

Again, I have assumed that our skier moves with a steady average speed, so that $v_n = v$. Substituting in equation (N13.6) from (N13.3) and (N13.4) we obtain, finally,

$$v = \frac{fd}{\Gamma} - \frac{\mu g(1 - \Gamma)}{2f}. \tag{N13.7}$$

Equation (N13.7) tells us that the skier's speed increases with increasing stride frequency (the faster she poles, the faster she goes), and that her speed increases with increasing cycle length. Both of these predictions are borne out by observation.[7] From the form of equation (N13.7) we see that speed also varies with duty factor, in such a way that it is a minimum for some value between $\Gamma = 0$ and $\Gamma = 1$. The skier doesn't want to minimize her speed—quite the opposite. So, this

7. Data from the women's 30-km event at the 1994 Winter Olympics in Lillehammer, reported in Rusko (2003), show that speed increases with stride frequency and cycle length.

equation tells us that skiers will tend to apply their poles to the ground very frequently (Γ near 1) or very infrequently (Γ near 0).

I leave it as an exercise for the reader (an exercise that is a lot easier than poling across snowfields, I assure you) to show that if the surface is inclined at a small angle a, the above analysis is still valid if we replace friction coefficient μ by $\mu + a$.

NOTE 14. FLYING DOWNHILL

Over what distance will a downhill skier be airborne? Say a skier takes off from level ground with an initial speed of v_0. (See fig. 6.2a for the hill geometry that I have in mind.) We will assume a shallow slope, so that the deceleration felt by the skier, due to drag, as she flies through the air is approximately $\ddot{x} \approx -bv^2$, which integrates to

$$x(t) = \frac{1}{b} \ln(1 + v_0 bt). \tag{N14.1}$$

Our skier is airborne for a time interval t_0, where

$$h = d \sin a = \tfrac{1}{2}gt_0^2. \tag{N14.2}$$

Figure 6.2a provides the notation for this equation. Solve equation (N14.2) for t_0 and substitute in equation (N14.1) to obtain

$$bd \cos a \approx \ln\left(1 + v_0 b \sqrt{\frac{2d}{g} \sin a}\right). \tag{N14.3}$$

In equation (N14.3) we have also substituted $x(t_0) = d \cos a$. We need to solve equation (N14.3) for the distance d, which is the distance over which our skier is airborne. Unfortunately, this equation is not solvable algebraically, so we must resort to number-crunching. Assuming that the drag factor $b = 0.0015$ m^{-1}, the slope angle $a = 20°$, and the initial speed $v_0 = 28$ ms^{-1} (all typical for a downhill skier), we find that the flight distance is $d = 57$ m.

We can obtain an *approximate* algebraic equation for d from (N14.3) if we assume that the aerodynamic drag is small, so that

$$d \approx \frac{2v_0^2}{g} \frac{\sin a}{\cos^2 a}. \tag{N14.4}$$

The approximation is quite crude (for example, for the parameter values we have chosen, it leads to $d = 62$ m, which is nearly 9% too high). However, it serves to show how flight length depends upon the other parameters. Thus, flight length increases as the slope gets steeper (to be expected) and as launch speed increases (ditto). From equation (N14.2) it also provides us with an approximate value for flight duration:

$$t_0 \approx \frac{2v_0}{g} \tan a, \tag{N14.5}$$

which yields 2 seconds for our chosen parameters.

NOTE 15. JUMP OR SLIDE?

Given a choice, should a skier jump or slide down a slope? Here, we calculate the time taken to slide down the slope and the time taken to slide over a level section of the course and then jump down the slope, such that the same distance is covered in both cases. (The notation used here is shown in fig. 6.2b.) Let us assume that two skiers start at the same place and with the same speed. One of them opts for the jump and the other for the continuous slide. We will see who takes the shortest time to get down the slope.

For this calculation, we can assume that aerodynamic drag will influence both skiers—the one who jumps and the one who slides—to about the same degree; therefore, for our purposes of estimating the time *difference*, we can ignore drag. We will also assume (to keep the math simple) that the friction coefficient is just enough so that the speed of a skier sliding down the slope is constant. That is, she neither slows down nor speeds up as she descends the slope. For this case, the sliding friction coefficient μ and the slope angle a are related via

$$\mu = \tan a. \tag{N15.1}$$

For the skier who elects to jump, her speed at the end of the level section (of fig. 6.2b) is $v_1 = v_0 - \mu g t_0$, where v_0 is the initial speed and t_0 is the time taken to traverse this level ground. The level ground is of length $x_0 = v_0 t_0 - \frac{1}{2}\mu g t_0^2$. Thus,

$$v_1 = \sqrt{v_0^2 - 2\mu g x_0}, \tag{N15.2}$$

$$t_0 = \frac{v_0 - v_1}{\mu g}. \tag{N15.3}$$

In these equations, μ is the sliding friction coefficient. During the flight stage there is no sliding friction, of course, and so the horizontal and vertical equations of motion lead to

$$x = v_1 t, \qquad z = h - \tfrac{1}{2} g t^2. \tag{N15.4}$$

The flight stage ends when the trajectory intersects the slope, that is, when x and z of equations (N15.4) satisfy

$$\tan a = -z/x. \tag{N15.5}$$

From equations (N15.1)–(N15.5) we see that the flight duration is given by

$$t_1 = \frac{1}{g} \left[\mu v_1 + \sqrt{\mu^2 v_1^2 + 2gh} \right]. \tag{N15.6}$$

So the total time it takes a skier to traverse the level surface and fly through the air is $t_0 + t_1$, obtained from equations (N15.3) and (N15.6).

For the skier who opts to slide all the way, the same horizontal distance is covered in a time

$$t_s = \frac{x_0 + v_1 t_1}{v_0 \cos a}. \tag{N15.7}$$

This equation is easy to derive from figure 6.2b and our assumption that slide speed is constant.

Now we can calculate the difference in time—call it τ.

$$\tau = t_0 + t_1 - t_s. \tag{N15.8}$$

The plots shown in figure 6.4 were drawn using equation (N15.8).

NOTE 16. JUMP START

In this note we will see why snowboarders choose to jump out of the trough that they encounter immediately after the start of a snowboard-cross race. Consider the simple trough of figure 6.8b, with the notation shown. We assume that initial speed v_0 is just sufficient for a snowboarder to slide through the trough, emerging at the far side with zero speed. Thus,

$$\tfrac{1}{2} m v_0^2 = 2\mu mgs, \tag{N16.1}$$

where $2s$ is the distance through the trough ($s = aR$). Equation (N16.1) says that the snowboarder's initial energy equals the work done in overcoming sliding friction. (We can neglect drag at the low speeds that apply here.) Now consider the snowboarder's energy at the start and at the bottom of the trough:

$$\tfrac{1}{2}mv_0^2 + mgh = \tfrac{1}{2}mv^2 + \mu mgs. \tag{N16.2}$$

Equation (N16.2) can be solved for the speed v at the bottom of the trough:

$$v = \sqrt{2gR(1 - \cos a + \mu a)}. \tag{N16.3}$$

Let us assume that the rider starts her jump at the bottom of the trough. She launches herself off the surface with a vertical speed V. It is not hard to show that (if her vertical acceleration was constant and brief) the jump slows her forward speed from v to $v_x = v - \mu V$. The vertical speed must be enough to get her out of the trough: if $V = \sqrt{2gh}$, then that will just be enough. Now look at the energy at the bottom of the trough and at the top, after the jump:

$$\tfrac{1}{2}mv_x^2 = mgh + \tfrac{1}{2}mv_1^2. \tag{N16.4}$$

We can solve equation (N16.4) for the rider's speed, v_1, as she emerges from the trough:

$$v_1 = \sqrt{2\mu gR(a - \sin a)}. \tag{N16.5}$$

This speed is greater than zero, and so the rider benefits from jumping.

There is a restriction imposed on this result, however. At the start of the jump, the snowboarder's vertical speed, V, and horizontal speed, v_x, must be enough to reach the top of the trough. Further calculations show that this requirement imposes a restriction on the shape of the trough:

$$3 - 5\cos a + 4\mu(a - \sin a) \geq 0. \tag{N16.6}$$

If the inequality (N16.6) is satisfied, then the jump works; otherwise, it doesn't. For a realistic value of the sliding friction coefficient $\mu = 0.15$, equation (N16.6) tells us that the trough angle a must exceed 52°—so the trough has to be steep-sided. Another practical restriction is that the trough height should be limited to the maximum that a rider can jump—say $h = 0.8$ m. This limits the trough radius to a maximum of about 2 m. If the trough is deeper (as for the track at Whistler), the jump must begin higher up the slope, nearer the end of the trough.

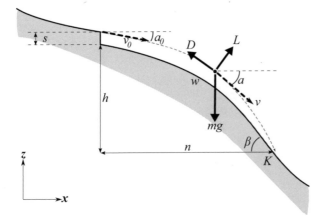

Figure N17. Ski jumper and ski hill. The jumper has an initial speed of v_0 and a take-off angle of a_0. When he is airborne, his speed is v and his direction angle is a. Bold arrows show the lift force, L; the drag force. D; and weight, mg. The take-off height is s, the hill height is h, and the horizontal jump distance to the K-point is n. The hill slope at the K-point is β.

NOTE 17. SKI JUMPING

To set up and solve the equations of motion for a simplified ski-jumping scenario, imagine the jumper to be a single point subjected to aerodynamic forces with constant lift and drag coefficients. The ski hill is shown in figure N17. The equations of ski-jumper horizontal and vertical motion are

$$\ddot{x} = b_L v^2 \sin a - b_D v^2 \cos a, \tag{N17.1}$$

$$\ddot{z} = b_L v^2 \cos a + b_D v^2 \sin a - g, \tag{N17.2}$$

where the skier's speed is $v = \sqrt{\dot{x}^2 + \dot{z}^2}$. There is no general analytic (that is to say, algebraic) solution to these equations, but they are amenable to numerical integration (that is to say, number-crunching). A realistic choice for the take-off speed is $v_0 = 25$ ms^{-1}, or about 56 mph. If the inrun angle is 11°, and the skier leaps upward with an initial lift speed of between 2.5 and 3 ms$_{-1}$ (a reasonable choice),[8] then his take-off angle is $a_0 = 4°$–5°. From figure 6.13 we see that a reasonable choice for lift-to-drag ratio (confirmed by wind-tunnel measurements) is $L/D = 1.1$, so that $b_L = 1.1 b_D$. Also, data show that the ratio of drag

8. If a ski jumper standing still on level ground is able to jump to a height of 0.45 m (18 inches), his vertical take-off speed (which we are calling initial lift speed) is 3 ms^{-1}.

force to the weight of a ski jumper is typically $D/mg = 0.4$ (see Müller 2009). Thus, $b_D = 0.4g/v_0^2$. So now we have fixed all the parameters needed to solve equations (N17.1) and (N17.2).

But what about the hill? We need to specify the hill shape so that we know when the jumper lands. To ensure that it is similar to the ski jumper's trajectory, so that the hill is never very far below him (for safety reasons), let us say that the hill shape has been chosen so that it is the shape of the trajectory of a ski-jumper with an initial speed of 26 ms^{-1}, all other parameters being equal. You can see that this will be a somewhat flatter shape than the actual trajectory, and so the two will intersect. We choose this intersection point to be the K-point. So, our hill has been designed for jumpers with specified lift and drag characteristics, and assuming a certain take-off speed. Let us also say that the shape of the hill beyond the K-point is straight—a uniform slope. This will ensure that any unusually fast jumpers also land not too far from the K-point. At even longer distances, of course, the slope levels off.

We now specify a typical value of $s = 3$ m for the take-off height (see fig. N17) and we are good to go. Results are plotted in figure 6.14.

BIBLIOGRAPHY

Albert, J., and R. H. Koning, eds. 2007. *Statistical Thinking in Sports*. Boca Raton, FL: Chapman and Hall/CRC Press.

Armenti, A. 1984. "How can a downhill skier move faster than a skydiver?" *The Physics Teacher* 22:109–10.

Avallone, E. A., T. Baumeister, and A. Sadegh. 2006. *Marks' Standard Handbook for Mechanical Engineers*. New York: McGraw-Hill.

Böhm, H., C. Schwiewagner, and V. Senner. 2009. "Simulation of puck flight to determine spectator safety for various ice hockey board heights." *Sports Engineering* 10:75–85.

Britannica. 1998. *Encyclopaedia Britannica*. CD 98 Standard Edition.

Castaldi, C. R., and E. F. Hoerner. 1989. *Safety in Ice Hockey*. West Conshohocken, PA: ASTM International.

Chang, K. 2006. "Explaining ice: The answers are slippery." *New York Times*, Science sec., Feb. 21.

Chapman, R. F., J. L Stickford, and B. D. LeVine. 2010. "Altitude training considerations for the winter sport athlete." *Experimental Physiology* 95:411–21.

Colbeck, S. C. 1995. "Pressure melting and ice skating." *American Journal of Physics* 63:888–90.

———. 1996. "A review of the friction in snow." In *Physics of Sliding Friction*, ed. B. N. J. Persson and E. Tosatti, 275–91. New York: Springer.

Colbeck, S. C., I. Najarian, and H. B. Smith. 1997. "Sliding temperatures of ice skates." *American Journal of Physics* 65:488–92.

Colbeck, S. C., and D. K. Perovich. 2004. "Temperature effects of black versus white polyethylene bases for snow skis." *Cold Regions Science and Technology* 39:33–38.

Copley-Graves, L. 1992. *Figure Skating History: The Evolution of Dance on Ice*. Columbus, OH: Platoro Press.

de Koning, J. J., et al. 1989. "Mechanical aspects of the sprint start in Olympics speed skating." *International Journal of Sports Biomechanics* 5:151–68.

———. 1995. "The start in speed skating: From running to gliding." *Medicine and Science in Sports and Exercise* 27:1703–8.

de Koning, J. J., G. de Groot, and G. J. van Ingen Schenau. 1992. "Ice friction during speed skating." *Journal of Biomechanics* 25:565–71.

de Mestre, N. 1990. *The Mathematics of Projectiles in Sport*. Cambridge: Cambridge University Press.

Denny, M. 2002. "Curling rock dynamics: Towards a realistic model." *Canadian Journal of Physics* 80:1005–14.

———. 2006. Comment on "On the motion of an ice hockey puck" by K. Voyenli and E. Eriksen. *American Journal of Physics* 74:554–56.

———. 2009. *Float Your Boat! The Evolution and Science of Sailing*. Baltimore: Johns Hopkins University Press.

Douglas, J. F., and R. D. Matthews. 1996. *Solving Problems in Fluid Mechanics*. Harlow, UK: Longman.

Evans, D. C. B., J. F. Nye, and K. J. Cheeseman. 1976. "The kinetic friction of ice." *Proceedings of the Royal Society of London A* 347:493–512.

Federolf, P., et al. 2006. "Deformation of snow during a carved ski turn." *Cold Regions Science and Technology* 46:69–77.

Garrett, W. E., and D. T. Kirkendall, eds. 1999. *Exercise and Sport Science*. Philadelphia: Lippincott Williams and Wilkins.

Gilenstam, K., K. Henriksson-Larsén, and K. Thorsen. 2009. "Influence of stick stiffness and puck weight on puck velocity during slap shots in women's ice hockey." *Sports Engineering* 11:103–7.

Goff, J. E. 2010. *Gold Medal Physics*. Baltimore: Johns Hopkins University Press.

Haché, A. 2002. *The Physics of Hockey*. Baltimore: Johns Hopkins University Press.

Howe, J. 1983. *Ski Mechanics*. Fort Collins, CO: Poudre Canyon Press.

Hubbard, M. 2002. "Recreating the Cresta Run." *Physics World*, Feb., 21–22.

IOC (International Olympics Committee). 2002. "Salt Lake City 2002 Olympic Winter Games Global Television Report." Prepared by Sports Marketing Surveys Ltd.

———. 2010. "Bobsleigh, luge and skeleton: From alpine traditions to specialist strategies." http://doc.rero.ch/lm.php?url=1000,44,38,20100614112951-HX/Bobsleigh_Luge_and_Skeleton_from_Alpine_Traditions_to_Specialist_Strategies_-_2002.pdf, accessed Dec. 6, 2010.

Jobse, H., et al. 1990. "Measurement of push-off force and ice friction during speed skating." *Journal of Applied Biomechanics* 6: 92–100.

Johnston, G. W. 1981. "The dynamics of a curling stone." *Canadian Aeronautical and Space Journal* 27:144–60.

Judd, R. C. 2009. *The Winter Olympics: An Insider's Guide to the Legends, Lore, and Events of the Games*. Seattle: Mountaineers Books.

Kaman, G., ed. 2001. *Foundations of Exercise Science*. Philadelphia: Lippincott Williams and Wilkins.

Kaps, P., W. Nachbauer, and M. Mössner. 1996. "Determination of kinetic friction and drag area in alpine skiing." In *Skiing Trauma and Safety*, ed. C. D. Mote et al., 10:165–77. ASTM STP 1266. West Conshohocken, PA: American Society for Testing and Materials.

Karydas, Thanos. 2008. "Breaking the Code: Understanding Wax Technology." Presentation at FIS International Coaches Forum, Naoussa, Greece, April 6, 2008. Online at www.bondars.lv/1_ALPINE/LITERATURA/2008_04_3-5_ SEMINARS-PROGRESS_ZIEMAS_SPORTU_TRENESANA/2008_07_23_ Sleposanas_invertara_sagatavosana.pdf.

Komi, P., ed. 2002. *Strength and Power in Sport: Olympic Encyclopedia of Sports Medicine*. 2nd ed. Oxford, UK: Wiley-Blackwell.

Lazar, M. A. 2003. *Let's Review Physics: The Physical Setting*. Hauppauge, NY: Barron's Educational Series.

Leino, A. H., E. Spring, and H. Suominen. 1983. "Methods for the simultaneous determination of air resistance to a skier and the coefficient of friction of his skis on the snow." *Wear* 86:101–4.

Lewis, O. 2006. "Aerodynamic analysis of 2-man bobsleigh." M.S. thesis, Delft University of Technology, The Netherlands. Available at www.lr.tudelft.nl/ live/pagina.jsp?id=d4d3851f-4916-4c6e-af53-31f35b703469andlang=enand binary=/doc/2006_1_15.pdf.

Libbrecht, K. G. 2005. "The physics of snow crystals." *Reports on Progress in Physics* 68:855–95.

Lomond, K. V., R. A. Turcotte, and D. J. Pearsall. 2007. "Three-dimensional analysis of blade contact in an ice hockey slap shot, in relation to player skill." *Sports Engineering* 10:87–100.

Luhtanen, P. 1996. "Wind tunnel measurements in ski jumpers and simulation of the jumps, Thunder Bay Hill K90." In *Proceedings of the XIIIth International Symposium of ISBS*, ed. T. Bauer, 240–45. Thunder Bay, ON: Lakehead University.

Luhtanen, P., J. Kivekäs, and M. Pulli. 2000. "Influence of a ski-jumper model, skis and suits on aerodynamical characteristics in ski jumping." In *Proceedings of XVIII International Symposium on Biomechanics in Sports*, ed. Y. Hong and D. P. Johns, 2:541–44. Hong Kong: Chinese University of Hong Kong.

Maryniak, J., E. Ladyzynska-Kozdras, and S. Tomczak. 2009. "Configurations of the Graf-Boklev (V-style) ski jumper model and aerodynamic parameters in a wind tunnel." *Human Movement* 10:130–36.

Massey, B. S. 1989. *Mechanics of Fluids*. London: Chapman and Hall.

Microsoft. 2005. *Encarta Encyclopedia*. Standard Edition. 2005.

Miller, P., et al. 2006. "Development of a prototype that measures the coefficient of friction between skis and snow." In *Engineering of Sport 6*, vol. 1, *Developments for Sports*, ed. E. Moritz and S. Haake, 305–10. New York: Springer.

Mills, A. 2008. "The coefficient of friction, particularly of ice." *Physics Education* 43:392–95.

Motallebi, F., E. Avital, and P. Dabnichki. 2002. "On the aerodynamics of the two-man bobsleigh." In *The Engineering of Sport 4*, 297–306. Hoboken, NJ: Wiley.

Müller, W. 2009. "Determinants of ski jump performance and implications for health, safety and fairness." *Sports Medicine* 39:85–106.

Müller, W., and B. Schmölzer. 2002. "The new jumping hill in Innsbruck: Designed by means of flight path simulations." In *Proceedings of the IVth World Congress of Biomechanics*. Calgary, AB: University of Calgary. Online at http://hpr.uni-graz.at/research_abstracts_download.php?id=21.

Nachbauer, W., P. Kaps, and M. Mössner. 1992. "Determination of kinetic friction in downhill skiing." In *Proceedings of the 8th Meeting of the European Society of Biomechanics*, 333–35. Rome. Online at http://sport1.uibk.ac.at/mm/publ/106—Nachbauer—1992—Determination_of_Kinetic_Friction_in_Down hill_Skiing.pdf.

New York Times. "To Learn about Skating, Study Sharpening." June 22, 2009.

Penner, A. R. 2001. "The physics of sliding cylinders and curling rocks." *American Journal of Physics* 69:332–39.

Pierce, A. 1997. "Shaped skis yearn to turn." *The Columbian*, Dec. 2.

Pinkus, O. 1987. "The Reynolds centennial: A brief history of the theory of hydrodynamic lubrication." *Journal of Tribology* 109:2–15.

Remizov, L. P. 1984. "Biomechanics of optimal flight in ski jumping." *Journal of Biomechanics* 17:167–71.

Reuters. 2006. "Scary Cesana track has its fans." Posted Feb. 16. www.redorbit.com/news/sports/393966/scary_cesana_track_has_its_fans.

———. 2009. "Housing market datapoint of the day." April 8. http://blogs.reuters.com/felix-salmon/2009/04/08/housing-market-datapoint-of-the-day.

Rusko, H., ed. 2003. *Handbook of Sports Medicine and Science: Cross Country Skiing*. Oxford: Blackwell.

Schmölzer, B., and W. Müller. 2002. "The importance of being light: Aerodynamic forces and weight in ski jumping." *Journal of Biomechanics* 35:1059–69.

Schulson, E. M. 1999. "The structure and mechanical behavior of ice." *JOM* 51:21–27.

Seo, K., I. Watanabe, and M. Murikami. 2004. "Aerodynamic force data for a V-style ski jumping flight." *Sports Engineering* 7:31–39.

Shegelski, M. R. A., and R. Niebergall. 1999. "The motion of rapidly rotating curling rocks." *Australian Journal of Physics* 52:1025–28.

Shegelski, M. R. A., R. Niebergall, and M. Walton. 1996. "The motion of a curling rock." *Canadian Journal of Physics* 74:663–70.

SMTC (St. Moritz Tobogganing Club). SMTC FAQs: "What is the Cresta Run?" St. Moritz Tobogganing Club Web site. www.cresta-run.com/html/smtc.cfm, accessed Dec. 6, 2010.

Smith, Gerald A. 2002. "Biomechanics of Cross Country Skiing." Draft chapter for *The Handbook of Sports Medicine and Science: Cross Country Skiing*, ed. Heikki Rusko. Oxford: Blackwell. Online at http://biomekanikk.nih.no/xchandbook.

Stefanyshyn, D. J., and B. M. Nigg. 2000. "Work and energy influenced by athletic equipment." In *Biomechanics and Biology of Movement*, ed. B. Nigg, B. Mac-Intosh, and J. Mester. Champaign, IL: Human Kinetics.

Symon, K. R. 1960. *Mechanics*. Reading, MA: Addison-Wesley.

van Valkenburg, P. 1988. "The aerodynamics of the bobsled." *Scientific American*, Feb., T10–14.

Voyenli, K., and E. Eriksen. 1985. "On the motion of an ice hockey puck." *American Journal of Physics* 53:1149–53.

Ward-Smith, A. J., and D. Clements. 1983. "Numerical evaluation of the flight mechanics and trajectory of a ski jumper." *Acta Applicandae Mathematicae* 1:301–14.

Watkins, J. 1998. *Introduction to Biomechanics of Sport and Exercise*. London: Churchill Livingstone.

Williams, C. 2001. *Science for Exercise and Sport*. New York: RoutledgeFalmer.

Witherell, W. 1988. *How the Racers Ski*. New York: W. W. Norton.

Zatsiorsky, V. M., ed. 2000. *Biomechanics in Sport: Performance Enhancement and Injury Prevention*, Vol. 9. Oxford, UK: Wiley-Blackwell.

INDEX